Earth Materials
and
Earth Processes
An Introduction

Earth Materials and Earth Processes

An Introduction

Third Edition

Lynn S. Fichter

James Madison University,
Harrisonburg, Virginia

G. Thomas Farmer

Earth and Environmental Sciences Division
National Laboratory, Los Alamos, New Mexico

Julia S. Clay

Blue Ridge Environmental Consultants
Harrisonburg, Virginia

Macmillan Publishing Company
New York
Collier Macmillan Canada
Toronto
Maxwell Macmillan International
New York Oxford Singapore Sydney

Macmillan Publishing Company
866 Third Avenue, New York, New York 10022

Collier Macmillan Canada, Inc.
1200 Eglinton Avenue, E.
Suite 200
Don Mills, Ontario, M3C 3N1

Library of Congress Cataloging-in-Publication Data

Fichter, Lynn S.
  Earth materials and earth processes: an introduction / Lynn S.
Fichter, George T. Farmer, Jr., Julia S. Clay. — 3rd ed.
    p.   cm.
  ISBN 0-02-337165-X
  1. Geology.   I. Farmer, George T.   II. Clay, Julia S.
III. Title.
QE26.2.F53   1991
550—dc20
Printing: 3 4 5 6 7 8

Preface

When we wrote the Second Edition to *Earth Materials and Earth Processes* we stated as our goal:

> . . . Knowing how to identify rocks and minerals, or even knowing a few facts about each is not in itself very interesting or important. Facts are worthwhile only if you know what to do with them. . . . We are . . . interested not only in the information contained in individual rocks and minerals, but also in their place in the world and the way geologists think about them, the special ways in which geologic events affect our view of the world, and the way in which problems in earth history are solved.

Since then as a primary goal the idea that the study of minerals, rocks, and the earth is not an end in itself but a medium for teaching students about how people do science, and how science defines and solves problems, has been strengthened. With the widely reported decline in American students' knowledge and understanding of science as compared to that of students in other industrialized nations, it is especially important today that young people not only be taught a science in enough detail to appreciate its larger theories, but also that they should practice solving scientific problems to understand the strategies of science in general.

Earth Material and Earth Processes is a college level physical geology laboratory manual specifically designed to teach students not only the theory and practice of geology but also the strategy of science. The manual is a complete text for those subjects customarily covered in introductory geology laboratories and is designed to complement any lecture text.

Most of the exercises for the third edition have been completely rewritten or revised. New identification keys have been designed, new illustrations prepared, and several new strategies and exercises for using earth materials and earth processes to learn about geology and science are integrated into the manual. A new exercise, "A Plate Tectonic Rock Cycle," which demonstrates how knowledge form all the previous mineral, rock, and structural geology exercises is integrated into a theory of the whole earth, has been developed.

Each exercise in the manual has been written and is organized with several levels and strategies of investigation providing the instructor with numerous choices and great flexibility in designing and running a laboratory, whether for the liberal arts student or the geology major. Even the simplest exercises in each laboratory are designed to challenged students with interesting problems while at the same time the students are enabled to develop critical observation and thinking skills and an appreciation of the scientific method. For the geology major, stimulating advanced problems are provided.

All the exercises in the manual are integrated from the most basic to the most advanced so that students can progress easily from level to level, and advanced students can move ahead to more challenging problems by building on what they have already learned. Preliminary chapters to be assigned as homework make a lab lecture unnecessary. The Preliminary chapter explains the terms, concepts, and theories needed to do the exercises. Each exercise contains detailed instructions that guide the student systematically through the lab, freeing instructors to concentrate on mentoring students.

An Instructor's Manual, designed especially to help the new instructor, is available at no cost by writing to the publisher. It contains suggestions for teaching introductory labs, and numerous hints, ideas, strategies, and support for those who are either old hands or just learning how to teach.

L. S. F.
G. T. F.
J. S. C.

Table of Contents

Introduction

A laboratory experience where the sole or main task is memorization, or "cookbook" experiments, or the "plug-and-chug" manipulation of numbers misses what is most essential about the laboratory. Holcomb and Morrison[1] express one purpose of the laboratory very well:

> The essence of learning in science is participation: doing and asking and making errors and learning from them. . . . The lab work is not primarily to train you in certain dexterous actions. It is only to give the genuine feel of that world which we can at best pallidly describe on paper in words or symbols. It is above all to give the sense of what is meant by *abstraction*. . . . we give up completeness by abstraction, but by it we gain knowledge and control.

An introductory laboratory to geology should do two things. The first is to bring firsthand familiarity with minerals and rocks and the ways geologists classify, study, and use them to interpret the earth. It is not genuinely possible to gain a feel for the earth if feldspar, basalt, schist, and breccia are only words or pictures. There must also be that tactile and visual familiarity which comes from handling the specimens, turning them over and over, studying them in ever more detail until the many levels of understanding begin to be revealed. The individual must work with the earth's materials until what is meant by critical observation is understood personally and data collection is revealed as not random or haphazard but as a selective process guided by theories. Study of the earth's materials must continue until the meaning behind classification systems is something not just memorized but understood because personal experience makes the logic of it evident.

The second thing the laboratory should do is give some insight into what is meant by a scientific study and a scientific conclusion. Studies of American students show that their knowledge of science is far below that of students in other industrial nations and has not improved significantly in recent years. Yet we live in a nation and world which is based on and is continuously being altered, for good or ill, by scientific advances. The future of the nation, and the well-being of the enterprise of science, is dependent on a citizenry who understands the process of science well enough to know what is, and what is not, a good scientific decision.

Specifically, you, the student, should come to understand that in science, just feeling, or even "knowing," that something is right is not enough; you must also logically demonstrate the point completely, from facts to conclusions, within a theoretical framework. You must also, with equal facility, be able to demonstrate why another solution is wrong within the theory, and do it in a way that would convince a skeptic. You must come to understand that science is not the random accumulation of knowledge but is the process of using hypotheses to make specific predictions which can be factually and logically demonstrated true, or false.

---

[1] D. F. Holcomb and Philip Morrison, *My Father's Watch: Aspects of the Physical World* (Englewood Cliffs, NJ: Prentice-Hall, 1974), p. 390

The revised sections of this lab manual were written with these goals continuously in mind. We have tried to write exercises that provide many different depths and strategies for exploring earth materials while at the same time being flexible and adaptable. We have included strategies to encourage the student to develop a scientific approach for studying the natural world, and questions and problems that will inculcate a facility for making scientific arguments.

We are aware of how difficult it can be to teach and learn these ideas about science. We are aware of how demanding it can be for the instructor to stay in touch with students who are struggling to learn a new subject and new ways of seeing and thinking. But we believe the earth is an endlessly fascinating laboratory which has much to teach us about the world and ourselves. We hope that the exercises in the manual will make it easier for geologists to use earth materials and earth processes to help students gain powerful insight into the methods, limits, and strengths of science. We also believe that the deeper engagement the students will have with geology through the exercises will more powerfully encourage them to have a lifelong fascination with and appreciation for the earth.

An Instructor's Manual available from the publisher at no charge to laboratory teachers and faculty provides additional strategies and suggestions and, for the first-time instructor, encouragement and hints on how to make a laboratory experience successful for both the instructor and student.

Earth Materials
and
Earth Processes
An Introduction

Minerals

PURPOSE

Why do we begin with the study of minerals, you may ask? Geology is the study of the earth, and minerals are the building blocks which make up the earth. Minerals are everywhere, and not just the obvious ones which have a geometric shape; the clay in the soil is also a mineral. Any discussion of the earth then must begin with an understanding of minerals. In later laboratories we will discuss how minerals combine to form the variety of rocks that occur on the earth. The purpose of this laboratory is twofold:

1. To teach you the fundamental skills of observation and hypothesis testing needed to do science and for the laboratories to follow.

2. To learn to identify common rock-forming minerals.

By the time you leave the laboratory today you should (1) be familiar with the physical properties of minerals, (2) be able to formulate and test hypotheses in order to identify unknown minerals, (3) be able to identify the minerals your instructor has for you, and (4) be intimate with the eight major igneous rock–forming minerals.

WHAT IS A MINERAL?

A mineral is a naturally occurring, usually inorganic, crystallized chemical element or compound, with a definite chemical composition which varies only within specific limits.

What does all this mean?

Although compounds can be produced in the lab that have many of the characteristics of a mineral, because they are not naturally occurring they are not minerals. Nearly all minerals are **inorganic,** that is, not produced by living organisms. Minerals are **crystallized,** that is, a solid with a definite internally ordered arrangement of atoms. Therefore, liquids and gases are not minerals. Because a mineral has a **definite chemical composition,** its composition can be expressed as a specific chemical formula. Quartz, for instance, is composed of silicon and oxygen and has the formula SiO_2. Other minerals such as garnet have more complex compositions. Garnet's formula is $A_3B_2(SiO_4)_3$, where the A and B represent a variety of elements such as Ca, Mg, Fe, Mn and so on. Although the exact composition can vary, the ratio is constant—3 moles of A for every 2 moles of B and always 3 molecules of SiO_4. Of course, exceptions can be found to every rule, but this definition will work for us in the laboratory.

ORGANIZATION

There are four parts to this laboratory, beginning with developing observation skills while learning mineral physical properties and then progressing through mineral identification. To progress to the next laboratory, you will need to master all four levels. A summary of the subjects in each part is as follows:

PART ONE—PHYSICAL PROPERTIES OF MINERALS   Explores the common physical properties of minerals and develops skills for testing and identifying physical properties.

PART TWO—MINERAL IDENTIFICATION   Applies the skills for identifying physical properties to the identification of minerals.

PART THREE—HYPOTHESIS TESTING   Develops a strategy for identifying difficult specimens and preparing to do well on mineral identification tests.

PART FOUR—IGNEOUS ROCK–FORMING MINERALS   Looks more closely at the eight minerals which comprise over 90% of the earth's crust and are essential for the identification of most igneous, sedimentary, and metamorphic rocks.

Part One

PHYSICAL PROPERTIES OF MINERALS

Physical properties are the characteristics minerals have because of their specific chemical compositions and crystalline structure. Because each mineral has a unique composition and structure, their physical properties are unique and can be used for identification. This is a lucky break for you. Instead of randomly memorizing trays of specimens, you need only learn a few physical properties by which nearly all common minerals can be identified.

Mineral identification is an integral part of a geologist's work. The method of identification used here depends on being able to determine the specific physical properties displayed by each mineral. The physical properties are **hardness, cleavage, fracture, luster,** and **streak,** and these properties are sufficient to identify most minerals. For a few minerals, additional physical properties such as the presence of **striations,** a reaction with dilute HCl acid, and **magnetism** will be necessary for identification.

INSTRUCTIONS

Note that throughout the manual instructions to do something are indicated by a square box followed by a number. The word "Instructions" will not appear with the boxes.

☐   (1)   Part one concerns learning to observe critically and using those observations to recognize different properties of nature. Along the way you will learn some properties needed to identify minerals. Take the time to look carefully at all the specimens and compare them for similarities and differences.

☐   (2)   Get a tray of minerals from your instructor. Work together in groups of two or three. You will learn faster and better if you can compare what you think you are observing with someone else's observations.

☐   (3)   Lay all the minerals out in front of you and leave enough room so you can arrange them in a row. You are going to learn to compare one mineral with another critically by first systematically testing mineral hardness and then cleavage and fracture.

TESTING
FOR HARDNESS

HARDNESS is a mineral's resistance to being scratched. Hardness is determined by how easily one mineral scratches another mineral, or a knife or a glass plate (H:5.5), or a penny (H:3.5), or your fingernail (H:2.5). Hardness is measured on the Mohs hardness scale from 1 (softest) to 10 (hardest).

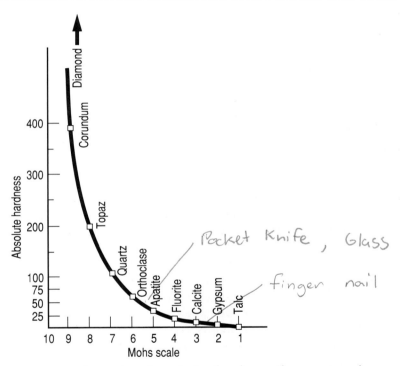

☐   (4)   Use the following steps to determine the relative hardness of your minerals:

STEP A   Try to scratch a piece of glass (H:5.5) with each mineral. Place the minerals softer than glass on the left and those harder than glass on the right. Have your instructor check your determinations.

> **OBSERVE:** Safety—place the glass plate flat on the table; do not hold it in your hand. Bear down firmly, but not so hard as to shatter the glass.

> **OBSERVE:** Always use a fresh, broken surface to make the scratch because many minerals alter to a material softer than the original mineral.

> **OBSERVE:** When one mineral is softer than another, the first may rub off on the second, leading you to think it is a scratch. This mark can be rubbed off. Your fingernail rubbed across a scratch should feel the nick in the glass.

STEP B Take the "softer than glass" minerals and try to scratch a penny (H:3.5). Place the minerals softer than a penny on the left and those harder than a penny on the right.

STEP C Take the "softer than a penny" minerals and try to scratch them with your fingernail (H:2.5). Place the minerals softer than your fingernail on the left and those harder than your fingernail on the right.

STEP D For the harder than glass (H:5.5) minerals, determine their relative hardness by scratching them against each other until the hardest are on the right and the progressively softer on the left.

> **OBSERVE:** Many minerals harder than glass are very close together in hardness, and you may have to test carefully and observe closely to arrange them correctly.

☐ (5) Your minerals should now be arranged from softest on the left to hardest on the right. Have your instructor check your arrangement of mineral hardnesses.

TESTING FOR CLEAVAGE AND FRACTURE

Cleavage and fracture are the ways a mineral breaks when a force is applied. How the mineral breaks is diagnostic of each mineral and can be used for identification.

Cleavage is the tendency for minerals to break along planes of atomic weakness. Because minerals are crystalline, these planes are regular and repeat at very closely spaced intervals. Minerals with cleavage always cleave in the same manner no matter how many times they are broken or how small the cleavage fragments are. [1]

Fracture occurs when the chemical bonds are of equal strength in all directions. When the mineral is broken, an irregular breakage results. A common type is **conchoidal fracture**, like the curved break of broken glass, but fracture may also be very rough and irregular.

Cleavage and fracture are the most difficult physical properties to identify and you might experience frustration until you learn to recognize them, but this is normal. Your lab instructor will probably have some good examples of cleavage and fracture that you can study. Some cleavage may be difficult to see in hand specimens. Ask your lab instructor to break some samples to demonstrate cleavage and fracture. Do not start breaking laboratory materials on your own. It is important to preserve specimens for the future students.

☐ (6) Use the following steps to determine the cleavage and fracture of your minerals.

---

[1] Next time you are eating, pour some salt on the table and observe the grains are cube-shaped. If you gently break a cube, the fragments will also be cube-shaped. No matter how small the fragments become they are always cube-shaped.

STEP A **Minerals which have fracture.** Separate in a pile to the right all the minerals which have fracture, that is, are broken in an irregular way, without flat, smooth, shiny surfaces. On the left, place all the other minerals with cleavage.

> **OBSERVE:** One way to recognize a cleavage face is to hold the mineral up to the light. Light reflects off a smooth cleavage face, but will appear dull if the face is just fractured.

> **OBSERVE:** Some specimens may seem to have flat cleavage surfaces, but you just can't decide if it is cleavage. In these cases look for another flat surface running the same way *within or on the other side of the specimen.* One cleavage surface is virtually always accompanied by others. For difficult specimens ask your instructor for help on what to observe.

STEP B **Minerals which have cleavage.** Cleavage is identified by the number of cleavage faces present and the angles between them. Place each mineral with cleavage over its appropriate drawing below.

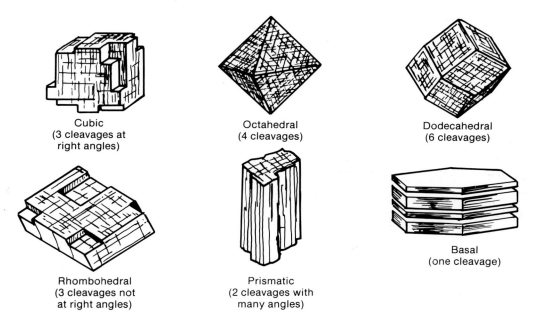

Cubic
(3 cleavages at
right angles)

Octahedral
(4 cleavages)

Dodecahedral
(6 cleavages)

Rhombohedral
(3 cleavages not
at right angles)

Prismatic
(2 cleavages with
many angles)

Basal
(one cleavage)

> **OBSERVE:** Minerals which grow crystal faces but have no cleavage. Some minerals grow crystal faces which look like cleavage faces, but the mineral may not have cleavage. Without breaking the specimen to see what happens it may be difficult to decide if the mineral does cleave.
>
> Carefully study the specimens with flat faces. If a flat face does ***not*** have other breaks running parallel to it within the specimen, or on a broken surface the mineral obviously fractures, the mineral probably does not have cleavage. Set these minerals in the fracture pile.

☐ (7) Ask your instructor to check your determinations of cleavage and fracture.

OTHER PHYSICAL PROPERTIES

Many physical properties other than hardness, cleavage, and fracture are found in minerals. Some are present in most or all minerals but are not very useful for mineral identification at our level. Some are found in only one or a few minerals and are diagnostic of the mineral.

Most of the other physical properties are not difficult to recognize and you will learn them while identifying minerals at Part Two. The properties used in the keys are described below.

STREAK is the color of a finely powdered mineral and is observed by rubbing the mineral on a porcelain **streak plate.**

OBSERVE: The streak plate has a hardness of about 7 and thus cannot be used on minerals of greater hardness.

OBSERVE: Some minerals will crumble when rubbed against a streak plate. Be sure the mineral is truly powdered.

LUSTER is the appearance of a mineral surface in reflected light. Many kinds of luster are recognized, but the following will be used.

1. **Metallic luster** looks like a metal and is of two types:
    A. Bright metallic luster reflects light like chrome, gold, steel or silver.
    B. Dull metallic luster is like rusted or weathered iron.
2. **Nonmetallic luster** does not look like a metal. It may be shiny, glassy, or dull.

OBSERVE: Bright metallic minerals tend to have a black or very dark streak.

OBSERVE: The streak of nonmetallic minerals is colorless or very light in color and probably not of much use in mineral identification.

☐ (8) Rub some of your "softer than glass" minerals against the streak plate and compare streak color with mineral color. Note especially the streak of bright metallic minerals.

STRIATIONS Some minerals have closely spaced fine lines on cleavage faces that look like straight record grooves. These lines result from the intergrowth of two crystals called **twinning**. Recognizing striations will be especially important in later labs.

MAGNETISM The magnet may pull at the mineral, but may not be able to suspend it. Magnetic minerals have a metallic luster.

COLOR Minerals come in many colors, and for *some* minerals color may be used to help identification. But color is *very* unreliable. Some minerals characteristically exhibit a variety of colors. *Never* trust color as the only basis for mineral identification.

┌─*Part Two*─────────────────────
MINERAL
IDENTIFICATION

You will have a set of unknown minerals to identify in the laboratory. These specimens have been collected and identified by someone who began his or her study of minerals in much the same way you are beginning yours.

☐   (1)   Get a tray of minerals from your instructor. Work together in groups of two or three.

☐   (2)   Select a mineral from the tray and write its number (if available) on the chart on pages 8–9. If the specimens are not numbered just arrange them in a row.

☐   (3)   Determine the physical properties of the mineral and list them on the chart.

☐   (4)   Use the "Key to the Identification of Minerals" on pages 10–11 and identify the mineral.

OBSERVE: Cleavage is an important discriminator for many minerals in the keys, but may be hard to see in specimens where the crystals are small or finely disseminated. Minerals which have cleavage but show it poorly are listed in the keys under *both* "cleavage" and "can't tell."

Other minerals may look like they have cleavage, but do not. These are in the "can't tell" category and it is indicated on the key as *not* having cleavage.

OBSERVE: You may expect your mineral specimens to be just one crystal with one set of cleavage faces, but some of your specimens are likely to be a large chunk made of hundreds or thousands of small crystals all jumbled together randomly. The small crystals have the same physical properties as a large crystal but the cleavage faces will be small and random, and difficult to see.

You may need a microscope to study these specimens. If you suspect a specimen is made of many smaller crystals ask your instructor and proceed with your analysis accordingly.

☐   (5)   After you have identified two or three minerals ask your instructor to check them before identifying the rest to make sure you are working correctly.

☐   (6)   When all the minerals are identified and you feel confident you know them,
      A.   Scramble all the minerals. Can you still identify them?

      B.   Exchange your tray for another and look at different specimens of the same minerals. Can you still identify them?

PHYSICAL PROPERTIES · SPECIMEN NUMBER · MINERAL NAME

HARDNESS	LUSTER	CLEAVAGE
STREAK	COLOR	OTHER

HARDNESS	LUSTER	CLEAVAGE
STREAK	COLOR	OTHER

HARDNESS	LUSTER	CLEAVAGE
STREAK	COLOR	OTHER

HARDNESS	LUSTER	CLEAVAGE
STREAK	COLOR	OTHER

HARDNESS	LUSTER	CLEAVAGE
STREAK	COLOR	OTHER

HARDNESS	LUSTER	CLEAVAGE
STREAK	COLOR	OTHER

HARDNESS	LUSTER	CLEAVAGE
STREAK	COLOR	OTHER

HARDNESS	LUSTER	CLEAVAGE
STREAK	COLOR	OTHER

HARDNESS	LUSTER	CLEAVAGE
STREAK	COLOR	OTHER

HARDNESS	LUSTER	CLEAVAGE
STREAK	COLOR	OTHER

PHYSICAL PROPERTIES

SPECIMEN NUMBER

MINERAL NAME

LUSTER	STREAK	HARDNESS
CLEAVAGE	COLOR	OTHER

LUSTER	STREAK	HARDNESS
CLEAVAGE	COLOR	OTHER

LUSTER	STREAK	HARDNESS
CLEAVAGE	COLOR	OTHER

LUSTER	STREAK	HARDNESS
CLEAVAGE	COLOR	OTHER

LUSTER	STREAK	HARDNESS
CLEAVAGE	COLOR	OTHER

LUSTER	STREAK	HARDNESS
CLEAVAGE	COLOR	OTHER

LUSTER	STREAK	HARDNESS
CLEAVAGE	COLOR	OTHER

LUSTER	STREAK	HARDNESS
CLEAVAGE	COLOR	OTHER

LUSTER	STREAK	HARDNESS
CLEAVAGE	COLOR	OTHER

LUSTER	STREAK	HARDNESS
CLEAVAGE	COLOR	OTHER

KEY TO MINERALS SOFTER THAN GLASS
Footnotes on Page 12

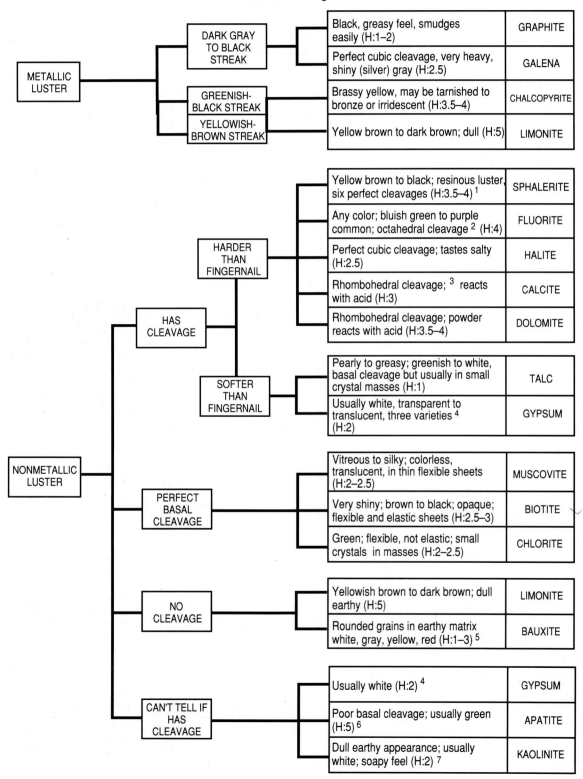

METALLIC LUSTER → **DARK GRAY TO BLACK STREAK**	Black, greasy feel, smudges easily (H:1–2)	GRAPHITE
	Perfect cubic cleavage, very heavy, shiny (silver) gray (H:2.5)	GALENA
GREENISH-BLACK STREAK	Brassy yellow, may be tarnished to bronze or irridescent (H:3.5–4)	CHALCOPYRITE
YELLOWISH-BROWN STREAK	Yellow brown to dark brown; dull (H:5)	LIMONITE

HARDER THAN FINGERNAIL	Yellow brown to black; resinous luster, six perfect cleavages (H:3.5–4) [1]	SPHALERITE
	Any color; bluish green to purple common; octahedral cleavage [2] (H:4)	FLUORITE
	Perfect cubic cleavage; tastes salty (H:2.5)	HALITE
	Rhombohedral cleavage; [3] reacts with acid (H:3)	CALCITE
	Rhombohedral cleavage; powder reacts with acid (H:3.5–4)	DOLOMITE
SOFTER THAN FINGERNAIL	Pearly to greasy; greenish to white, basal cleavage but usually in small crystal masses (H:1)	TALC
	Usually white, transparent to translucent, three varieties [4] (H:2)	GYPSUM

HAS CLEAVAGE (NONMETALLIC LUSTER)

PERFECT BASAL CLEAVAGE	Vitreous to silky; colorless, translucent, in thin flexible sheets (H:2–2.5)	MUSCOVITE
	Very shiny; brown to black; opaque; flexible and elastic sheets (H:2.5–3)	BIOTITE
	Green; flexible, not elastic; small crystals in masses (H:2–2.5)	CHLORITE

NO CLEAVAGE	Yellowish brown to dark brown; dull earthy (H:5)	LIMONITE
	Rounded grains in earthy matrix white, gray, yellow, red (H:1–3) [5]	BAUXITE

CAN'T TELL IF HAS CLEAVAGE	Usually white (H:2) [4]	GYPSUM
	Poor basal cleavage; usually green (H:5) [6]	APATITE
	Dull earthy appearance; usually white; soapy feel (H:2) [7]	KAOLINITE

6 Harder than 1-7
3 Harder than 6

KEY TO MINERALS THAT SCRATCH GLASS
Footnotes on Page 12

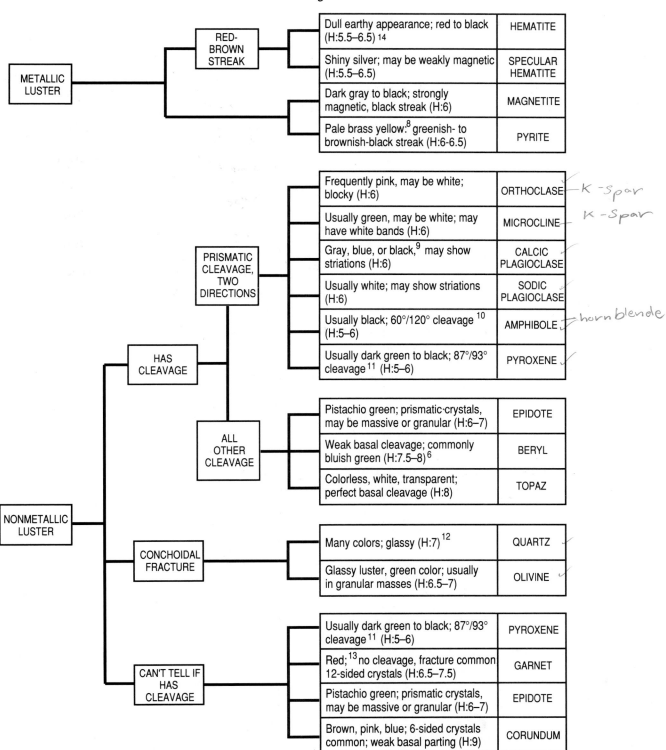

METALLIC LUSTER	RED-BROWN STREAK	Dull earthy appearance; red to black (H:5.5–6.5) [14]	HEMATITE
		Shiny silver; may be weakly magnetic (H:5.5–6.5)	SPECULAR HEMATITE
		Dark gray to black; strongly magnetic, black streak (H:6)	MAGNETITE
		Pale brass yellow;[8] greenish- to brownish-black streak (H:6-6.5)	PYRITE

NONMETALLIC LUSTER	HAS CLEAVAGE	PRISMATIC CLEAVAGE, TWO DIRECTIONS	Frequently pink, may be white; blocky (H:6)	ORTHOCLASE — K-spar
			Usually green, may be white; may have white bands (H:6)	MICROCLINE — K-spar
			Gray, blue, or black,[9] may show striations (H:6)	CALCIC PLAGIOCLASE
			Usually white; may show striations (H:6)	SODIC PLAGIOCLASE
			Usually black; 60°/120° cleavage [10] (H:5–6)	AMPHIBOLE — hornblende
			Usually dark green to black; 87°/93° cleavage [11] (H:5–6)	PYROXENE
		ALL OTHER CLEAVAGE	Pistachio green; prismatic crystals, may be massive or granular (H:6–7)	EPIDOTE
			Weak basal cleavage; commonly bluish green (H:7.5–8)[6]	BERYL
			Colorless, white, transparent; perfect basal cleavage (H:8)	TOPAZ
	CONCHOIDAL FRACTURE		Many colors; glassy (H:7)[12]	QUARTZ
			Glassy luster, green color; usually in granular masses (H:6.5–7)	OLIVINE
	CAN'T TELL IF HAS CLEAVAGE		Usually dark green to black; 87°/93° cleavage [11] (H:5–6)	PYROXENE
			Red;[13] no cleavage, fracture common 12-sided crystals (H:6.5–7.5)	GARNET
			Pistachio green; prismatic crystals, may be massive or granular (H:6–7)	EPIDOTE
			Brown, pink, blue; 6-sided crystals common; weak basal parting (H:9)	CORUNDUM

NOTES FOR MINERALS KEY
Softer than Glass

1. **Sphalerite**—powdered streak smells of rotten eggs; often in masses of smaller crystals.

2. **Fluorite**—because fluorite has perfect octahedral (4-directional) cleavage look for triangular facets and fractures within the crystal.

3. **Calcite**—Calcite forms perfect six-sided crystals, but such specimens are rare in introductory labs. White or colorless but many colors from contamination.

4. **Gypsum**—There are three varieties of gypsum. Satin spar is fibrous with a silky luster. Alabaster is fine-grained and massive. Selenite is colorless and transparent, sometimes in thick plates with basal cleavage. Rarely pink.

5. **Bauxite**—Frequently bauxite scratches glass due to impurities within the mineral. Don't be fooled.

6. **Apatite**—Apatite and beryl may be confused; however, beryl scratches glass.

7. **Kaolinite**—Resembles chalk but does not react with acid. Kaolinate has a soapy feel while chalk is gritty.

Minerals That Scratch Glass

8. **Pyrite**—Frequently forms cubes with striated faces. This is not cubic cleavage but crystal growth. Pyrite fractures when broken. Not to be confused with chalcopyrite which does not scratch glass. Sometimes called "fool's gold."

9. **Calcic plagioclase**—Most lab specimens are gray but pure calcic plagioclase may be white. Plagioclase of intermediate composition, called Labradorite, shows a brilliant iridescent of colors. Striations looking like fine, straight record grooves on cleavage faces are diagnostic—look for them.

10. **Amphibole**—Usually forms elongated crystals. Cleavage difficult to discern. Look for presence of "^" rooftop intersections of cleavage in end view. A crystal rotated in the light will show two "light flashes" coming off the crystal faces at less than 90° rotation. One light flash is very bright, the other less so. Because amphibole and pyroxene are so difficult to identify in hand specimen we will identify black, elongate shiny minerals as amphibole.

11. **Pyroxene**—Usually forms blocky crystals. Cleavage difficult to discern, especially in massive specimens. Because pyroxene and amphibole are so difficult to identify in hand specimen we will identify dark green, dull blocky minerals as pyroxene.

12. **Quartz**—Quartz comes in many colors and varieties. Rock crystal is usually colorless with distinct six-sided crystals. Amethyst-violet quartz. Rose quartz–coarsely crystalline, usually without crystal form. Smokey quartz–smokey yellow to almost black. Milky quartz–milky white. Quartz is one of the most common minerals encountered in rocks and you will have to learn to recognize it in all its varieties.

13. **Garnet**—Frequently red, however, may be brown, yellow, white, green, or black. Fracture can often be confused for cleavage; be cautious.

14. **Hematite**—Although hematite does form crystals most of the time it is powdery or finely disseminated. Some lab specimens are "oolitic," that is, forming small spherical grains.

┌─*Part Three*─────────────────────────┐
HYPOTHESIS
└────────── TESTING──────────────┘

There are two basic ways to learn minerals for an exam. The "brute force" method, "I'll just stare at the specimens so long maybe I'll recognize them on the exam," which produces wrong answers more times than right, and the "hypothesis testing" method, which produces right answers most of the time.

HYPOTHESIS TESTING MINERALS YOU BELIEVE YOU KNOW

Most minerals once identified are easy to remember. A problem you may have is seeing another specimen which while the same mineral, differs in some superficial way, like color. Therefore, even though you know these minerals, each time you pick up a specimen to study it say to yourself:

> Hypothesis: This mineral is galena, therefore . . . . Test: It will feel really
> heavy, have perfect cubic cleavage, and have a dull gray streak.

Then test the mineral for these physical properties. If the specimen has these required physical properties, then the identification is probably correct, even if the specimen is somewhat unusual.

If the specimen is missing one or more of these important physical properties, then your hypothesis is wrong. Do not try to identify the mineral as galena; that answer will be wrong. You will have to find another mineral name to hypothesize and test.

An example of how the hypothesis test method may prevent you from making a mistake is the mineral calcite. Calcite has rhombohedral cleavage, but when calcite grows in an open space ,it forms a six-sided crystal with a pointed end looking superficially like a quartz crystal. If you were to see the six-sided calcite crystal and say:

> Hypothesis: This mineral is quartz, therefore. . . . Test: It will be harder than
> glass, and fracture but not cleave.

Then, running just one of the tests (hardness or fracture, not cleavage) tells you immediately the specimen is not quartz. Then you must observe, again, this time more carefully, to form another hypothesis to test. You must admit, it is better to know that you do not know what a mineral is than to believe you know what it is and get it wrong on a test. To know you do not know gives you a second chance to get it right.

The habit of establishing a hypothesis (what mineral you think it is) and then listing the characters you will observe if the hypothesis is valid trains your mind to think systematically. Study and practice this way so that during an exam the hypothesis-test strategy will be automatic, and new or unusual specimens will not fool you.

HYPOTHESIS TESTING MINERALS WHICH MAY CONFUSE YOU

What do you do with all those clear to white minerals you have to learn? How do you distinguish among halite, calcite, dolomite, gypsum, talc, colorless fluorite, kaolinite, quartz, sodic plagioclase feldspar, and white orthoclase? Some of these are easy, such as halite, which tastes salty and has perfect cubic cleavage. Among the others there are some obvious distinctions which allow you to separate the minerals into groups, such as harder or softer than glass and has cleavage–does not have cleavage.

On the other hand, how do you tell the difference between gypsum, talc, and kaolinite, all three of which are easily scratched by your fingernail? Kaolinite and talc are similar because they are both clays. However, in hand specimen kaolinite feels powdery and yields a powder when scratched. Talc, on the other hand, yields small, perfect, basal cleavage fragments when you scratch it. Gypsum does not feel slippery or greasy and also yields a powder when you scratch it. All this information is useful to you, but it needs to be well organized. Below is a decision tree which uses these characteristics to differentiate among these similar minerals.

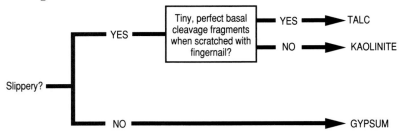

(1) If available, pick out and lay in front of you the following sets of minerals. List the most important physical properties of the minerals in each set. You may not use color because many of these minerals come in many colors, including colorless.
  A. Fluorite, calcite, dolomite
  B. Calcic plagioclase, pyroxene, amphibole
  C. Olivine, epidote, apatite, chlorite, beryl
  D. Pyrite, chalcopyrite, magnetite, specular hematite, earthy (red) hematite

(2) On a separate piece of paper draw a decision tree like the one above for talc, kaolinite, and gypsum which will allow you to distinguish among the minerals in these sets.
  When finished, compare your decision trees with ours at the back of the chapter. Please do not look at our decision trees before you draw your own. If you are unable to draw a decision tree which works then you should get help.

> **OBSERVE:** Decision trees have many formats. Your decision trees may not look like ours, but if they work that is all that is important.

(3) Now pick out any other minerals which confuse you. Ask, "What two or three tests will separate these minerals?" Look up the properties of the minerals, run the tests, and develop a hypothesis-test strategy which *you* can use to separate and identify the specimens.

(4) Use this strategy for learning to identify the igneous, sedimentary, and metamorphic rocks in the following labs.

Part Four

IGNEOUS
ROCK–FORMING MINERALS

The eight igneous rock–forming minerals make up over 90% of the earth's crust. Thus, they are of considerable importance in geologic studies, and it is important that you be able to recognize and identify them in hand specimen, and especially under a microscope. If you make these observations now, the next laboratory will be easier, with less frustration and anxiety.

☐ (1) Place the minerals listed below in a pile in front of you.

A. Olivine	E. Calcic and sodic plagioclases
B. Pyroxene	F. Orthoclase
C. Amphibole	G. Muscovite
D. Biotite	H. Quartz

☐ (2) Now that you are skilled in observing physical properties you need to apply those skills under new conditions. Look at each of the eight rock-forming minerals, first in hand specimen and then under a microscope. Do the physical properties look the same under both conditions?

☐ (3) Compare back and forth under the microscope the following *pairs* of minerals.
   A.   Olivine with quartz
   B.   Pyroxene with amphibole
   C.   Biotite with amphibole
   D.   Quartz with sodic plagioclase
   E.   Calcic plagioclase with amphibole
   F.   Orthoclase with sodic plagioclase

   How similar or different do they appear? Are you sure you can accurately observe the physical properties which will allow you to recognize and identify them?

☐ (4) Have your partner place one of the igneous rock–forming minerals under the microscope without your seeing the hand specimen. How confident are you that you can identify it? In most igneous rocks (next lab), these minerals are only a few millimeters long and smaller, and you have to be able to identify them still!

NOTICE

The next laboratory on igneous rocks, as well as most labs throughout the manual, have a Preliminary reading. The Preliminary is to be read **before** coming to the laboratory. The Preliminary readings provide the theoretical background necessary to understand and do the lab exercises.

DECISION TREES FOR DISTINGUISHING AMONG MINERALS LISTED UNDER STEP 3, PAGE 13

Decision Tree for Talc, Kaolinite, Gypsum

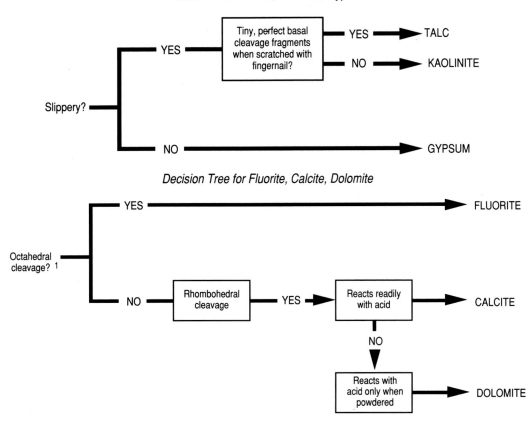

Decision Tree for Calcic Plagioclase Feldspar, Amphibole, Pyroxene

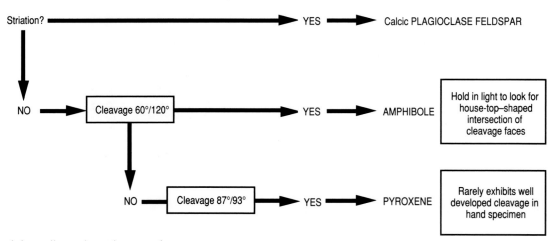

[1] Look for small triangles on faces or within specimen.

Decision Tree for Olivine, Epidote, Apatite, Chlorite, Beryl

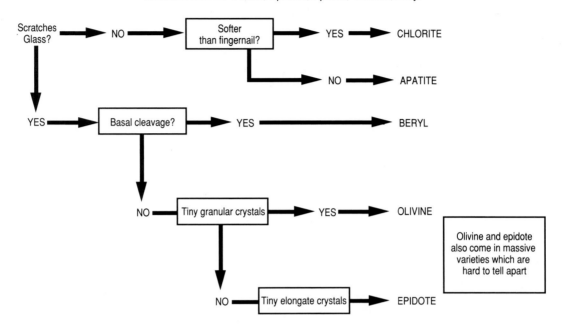

Decision Tree for Pyrite, Chalcopyrite, Magnetite, Specular Hematite, Oolitic Hematite

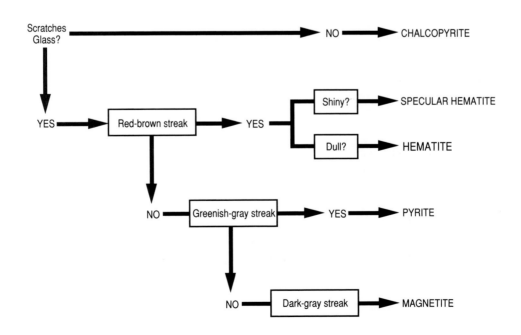

Preliminary to Igneous Rocks

PURPOSE

The igneous rock laboratory will build on the observation skills and knowledge of minerals you gained in the minerals lab. Upon completion of this "Preliminary," you will know that most igneous rocks are composed of the eight igneous rock–forming minerals. You will understand how a descriptive classification based on texture (size and shape of mineral grains) and composition leads to an understanding of the environment of formation (genesis) of igneous rocks.

This preliminary introduces some terms and concepts which will be necessary to understand the laboratory which follows. By the time you enter the next laboratory, you should:

1. Know how igneous rocks form and why they are important.

2. Know the different igneous rock textures—**glassy, crystalline, vesicular,** and **pyroclastic; phaneritic, aphanitic,** and **porphyritic;** and what is **groundmass** and a **phenocryst**—and how these relate to cooling history.

3. Know the minerals typical of **mafic, intermediate,** and **felsic** magmas and igneous rocks and how they affect rock color.

4. Have memorized **Bowen's Reaction Series** and know the temperature, color, density, viscosity, and composition trends within it.

INTRODUCTION

What Igneous Rocks Are Good For.   Both nongeologists and geologists alike are attracted to igneous rocks because of their beauty and durability. They are widely used for building and tombstones, and practically everyone has seen igneous rocks whether they know it or not. But igneous rocks are scientifically important because most of the earth is made of igneous rocks, and most of them have formed deep inside places we can never go to or see. Studying igneous rocks which are now exposed at the earth's surface allows us to begin to understand the earth and its processes.

To begin to understand the earth is to realize that each igneous rock forms only under a specific set of geological conditions. Describe the conditions and you can predict the rock which will result. Find the rock and you know what conditions existed at the time of its formation. To do this, you must understand the processes responsible for igneous rock texture and composition.

COOLING HISTORY AND IGNEOUS ROCK TEXTURE/COMPOSITION

All igneous rocks crystallize from **magma,** or molten rock. Magma is generated within the earth and can be **intrusive** (injected into the rocks of the earth) or **extrusive** (spilled out onto the earth's surface). Magma which is extruded becomes **lava.** The term "lava" applies to both the liquid and solid forms.

A crucial key to understanding igneous rocks is to understand that a magma of a particular composition can develop into more than one kind of igneous rock depending on cooling history. Cooling history affects the final rock in two very different ways. First, speed of cooling determines how large or small the crystals are, that is, the rock's **TEXTURE**. An intrusive magma which cools slowly, deep underground, and forms large crystals makes a rock with a different name from the same magma cooling quickly and forming microscopic crystals. This is true even if the magma compositions are identical. For example, a gabbro and a basalt can be identical in composition but look very different in hand specimen.

A second way cooling history affects an igneous rock is if a magma cools in the right way it is possible for it to become split, or fractionated.[1] In fractionation an original magma of one composition can become two rocks, each with a composition different from the original magma. In fact, the earth began with only a few kinds of magma and from them the great diversity of igneous rocks now known have evolved. This process will be important when learning how igneous rocks are interpreted and will be discussed in the lab *A Plate Tectonic Rock Cycle.*

Cooling Histories and Kinds of Texture   A lava is a disorganized mixture of many chemical elements and compounds. If it cools very rapidly crystals are unable to nucleate and grow before the lava solidifies. In this case no minerals are able to form; the lava has the same disorganized structure as a solid as it had as a liquid. The resultant rock is a **glass**. **Obsidian** is an example of a glass.

If the lava cools more slowly, the elements and compounds in the lava have time to come together and organize, and crystals form and grow. These crystals begin so tiny they cannot be seen without a microscope. **Crystalline** rocks (those composed of crystals, i.e., minerals) in which individual crystals are not visible have an **aphanitic** texture. **Basalt** is an example of aphanitic texture.

Magma which cools deep within the earth is insulated and, therefore, may cool very slowly. Crystals not only nucleate, but they have the time to grow large enough to be seen in hand specimen. This texture is called **phaneritic. Granite** is an example of phaneritic texture.

When magma cools extremely slowly, very-coarse-grained crystals grow (>1 cm). This texture is called **pegmatitic.** [2]

Two-Stage Cooling   Sometimes a magma begins cooling slowly at depth, but then is suddenly extruded onto the surface. In this case, the rock has both a phaneritic component (**phenocrysts**) resulting from the slow cooling and an aphanitic component (**groundmass**) formed from the rapid cooling. The resulting rock is a porphyry.

Pyroclastic Rocks   Magmas and lavas not only contain the materials to make minerals, they also contain gases and water (as steam), usually under great pressure. When a magma cools completely below ground the gases and water are gradually released from the magma and escape into the surrounding rock (called **country rock**) as the magma solidifies. If, however, the magma is ejected onto the earth's surface and the pressure is suddenly released, the gases and water escape quickly and explosively. If you have seen pictures of an erupting volcano and seen material being blasted into the air you have a sense of how violently the gases and water can be released. A variety of igneous rocks form from this process.

---

[1] Fractionation also occurs if an igneous rock is slowly heated until it begins to melt. Because it melts in stages the original rock can be separated into a liquid fraction and an unmelted residue. This process will be examined in a couple of later labs.

[2] Very-coarse-grained crystals up to many feet long may also grow if a lot of water is in the magma. Water moves the elements about quickly and easily, continuously supplying the material the crystal needs to grow.

A **vesicular rock** is usually an aphanitic or porphyritic rock which is full of vesicles, or holes. The vesicles are formed by expanding gas in the last stages of cooling.

Pumice is a light-colored vesicular, glassy rock. That is, it has cooled so quickly that minerals have not had time to form, and the gas has escaped quickly enough to froth the lava into a foam. Pumice is usually very lightweight because it is mostly holes.

Scoria is a dark-colored vesicular igneous rock. It is heavier than pumice and is usually not as glassy as pumice.

Tuff is a general term applied to all consolidated pyroclastic rocks. Some tuffs are **welded tuffs**, that is, glass-rich pyroclastic rocks welded together while still very hot. Frequently, these rocks look as if they are made of many smaller fragments and may be banded or streaked.

Other kinds of pyroclastic rocks also form, but the ones above are most commonly studied in geology laboratories. We will deal with each of these rocks in the classification systems developed in the laboratory. What you need to take with you from this is the fact that the texture of an igneous rock tells you a great deal about the cooling history of the rock.

TEXTURE	COOLING HISTORY	EXAMPLE
Glassy	Extremely rapid cooling; noncrystalline	Obsidian
Vesicular	Extremely rapid cooling with rapid gas escape forming bubbles in the rock	Pumice, scoria
Aphanitic	Fast cooling; microscopic crystal growth	Rhyolite, andesite, basalt
Phaneritic or pegmatitic	Very slow cooling; crystals grow to visible size; pegmatite crystals are >1 cm and can become meters in size	Granite, diorite, gabbro
Porphyritic	Two-stage cooling, one slow underground, the second rapid at the earth's surface.	Any aphanitic rock with the adjective porphyry
Pyroclastic	Explosive release of gases shattering the lava and sometimes welding the fragments together	Tuff; welded tuff

COLOR AND COMPOSITION IN IGNEOUS ROCKS

Igneous rocks are composed primarily of the eight rock-forming minerals (chart below). The rock-forming minerals are placed into two categories called **mafic** and **felsic.** Sometimes it will be necessary to talk of individual minerals, and sometimes to refer to the mafic and felsic minerals as a group:

IGNEOUS ROCK–FORMING MINERALS			
Mafic Minerals *(Rich in iron, and magnesium: generally dark colored)*		**Felsic Minerals** *(Rich in sodium, potassium, silica: generally light colored)*	
	Appearance in Hand Specimen		Appearance in Hand Specimen
Olivine	Pale green, glassy	Sodic plagioclase	White/light colored with striations
Pyroxene	Dark, dull greenish blocks	Orthoclase	Pink/white usual
Amphibole	Shiny black elongate crystals	Muscovite	White/brassy plates
Biotite	Shiny black plates	Quartz	Clear/glassy
Plagioclase[1]	Gray, with striations		

Recall from the last lab each of the minerals in the chart. There is a relationship between mineral color and the composition of igneous rocks which will be useful to you in the lab. Dark-colored rocks tend to be **mafic** (high in mafic minerals), such as olivine, pyroxene, and amphibole. Light-colored rocks tend to be **felsic** (high in feldspar and silica (e.g., quartz)).

Although it is possible for magmas/lavas to be virtually any composition, and thus to be any color, it turns out most magmas/lavas are one of only three kinds. This fact greatly simplifies your study of igneous rocks. Studying igneous rocks with this in mind cuts through a lot of confusion you could experience. The kinds of magma/lava are in the chart below:

Kind of Magma	**Rock Color**	**Typical Minerals Formed from Crystallization**
Mafic	Dark	Olivine, pyroxene, amphibole, calcic or intermediate plagioclase
Intermediate	Intermediate	Amphibole, intermediate plagioclase
Felsic	Light	Quartz, orthoclase, sodic plagioclase, biotite

You will find color a useful first approximation for identifying igneous rocks and our first classification is based on it. But the only accurate method for rock identification is to determine the

[1] Calcic plagioclase is not a mafic mineral, although it is often associated with them.

content and relative abundance of minerals. Because crystals in an aphanitic rock are microscopic, these rocks must be studied in **thin section** (paper-thin slices of the rock) with a polarizing microscope; for this reason the laboratory will not deal with composition in aphanitic rocks.

For phaneritic rocks, however, you will hone your mineral identification skills by estimating relative abundances of quartz, feldspars, and mafic minerals and use these estimates to name the rock.

IGNEOUS ROCK CLASSIFICATION

The challenge of any classification is always to devise a system which conveys as much information as possible as efficiently as possible. Igneous petrologists [1] classify igneous rocks not only because they want to emphasize the *differences* between rocks in a particular study area but also because they want to emphasize the *similarities* among rocks from one area to another. This is because minor differences in composition among igneous rocks in one region can mean significant geological changes, while detailed similarities in widely separated rocks can indicate that they formed in the same way, even though all the other evidence is gone.

At a minimum a classification system should include information on texture and composition, but part of the difficulty with classifying igneous rocks is that nature does not provide any natural divisions. Igneous rock composition and texture vary continuously. Thus, all classifications are arbitrary. Because of this, geologists have devised many classifications systems to deal with igneous rocks. Some of these classifications are complex, dividing igneous rocks into many subdivisions. Yet virtually every geological study begins with a simple classification like the ones you will learn in the laboratory. From those simple classifications the more technical classifications can be readily understood.

The next laboratory introduces two levels of classification:
1. The first classification, based on TEXTURE (size of the crystals) and OVERALL COLOR of the rock, is the simplest. This classification is a first approximation and lumps igneous rocks into major categories.

2. The second classification is based on TEXTURE and COMPOSITION and is a more sophisticated approach. It requires being able to recognize individual minerals in the rock and estimating their abundance. It is done in hand specimen with a hand lens (10x) or low-power microscope.

A summary of the simplest classification of igneous rocks we use is in the chart on the next page. Note several relationships. First, igneous rock classification is related to the mafic, intermediate, and felsic magmas listed in the chart on page 22. Second, the minerals which can crystallize are directly related to the type of magma. Third, the minerals crystallizing in mafic magmas tend to be dark, leading to dark-colored rocks; the minerals crystallizing in felsic magmas tend to be light colored leading to light-colored rocks; the minerals crystallizing from intermediate magmas are mixed in color leading to intermediate-colored rocks.

---

[1] People who study the origin, occurrence, and history of igneous rocks.

MAGMA TYPE	ROCK/ MINERAL COLOR	ROCK NAMES		
		Slow Cooling, Coarse Grained	Fast Cooling, Fine Grained	Two-Stage Cooling, Coarse and Fine
Mafic	Dark crystals	**Gabbro**	**Basalt**	**Basalt porphyry**
Intermediate	Dark and light crystals mixed	**Diorite**	**Andesite**	**Andesite porphyry**
Felsic	Light crystals	**Granite**	**Rhyolite**	**Rhyolite porphyry**

In addition to the rock names in the chart are the glassy, vesicular, porphyritic, and pyroclastic varieties, some of which will be introduced in the laboratory.

BOWEN'S REACTION SERIES

Nineteenth-century geologists observed that among the rock-forming minerals certain pairs of minerals never occurred together, while other pairs frequently occurred together. For instance, calcic plagioclase does not occur with quartz, and muscovite is never found with olivine. On the other hand, the mineral pairs of quartz and orthoclase, and pyroxene and calcium plagioclase, and amphibole and plagioclase (of intermediate composition) commonly coexist. In 1928 N. L. Bowen summarized these relationships in a model known as **Bowen's Reaction Series.**

Bowen's Reaction Series is one of the most powerful models we have for understanding igneous rocks. It can be used to help describe igneous rocks and is an extremely powerful tool for interpreting the origin of igneous rocks. You can learn igneous rocks without knowing Bowen's Reaction Series, but the reaction series give a rhyme and reason to why igneous rocks are the way they are, and how they are related. It takes a little more time to understand igneous rocks with Bowen's Reaction Series, but then they make much more sense and are much easier to study and remember.

The reaction series describes how the eight rock-forming minerals are related to each other. The most basic relationship is temperature. Begin with an ideal magma and allow it to cool. All the minerals do not crystallize out at the same time. Olivine and calcic plagioclase feldspar crystallize first somewhere around 1400° C, resulting in a mixture of magma and crystals. As the temperature continues to drop, other minerals crystallize out of the magma until finally at 570° C the last mineral, quartz, crystallizes out; see figure on next page. At this point, there is no magma left.

Bowen's Reaction Series not only relates the rock-forming minerals by temperature but by many other physical properties. We are going to explore only those relationships which will help you learn to identify igneous rocks. In general, minerals at the top of Bowen's Reaction Series:

1. are dark in color (i.e., form mafic magmas), while those at the bottom are light in color (i.e., form felsic magmas). In the middle, then, are intermediate magmas, which have the mixture of minerals.

2. have a greater density than do those at the bottom.

3. are more fluid (less viscous) than are those at the bottom.

4. are rich in iron, magnesium, and calcium, while those at the bottom are rich in potassium and silica.

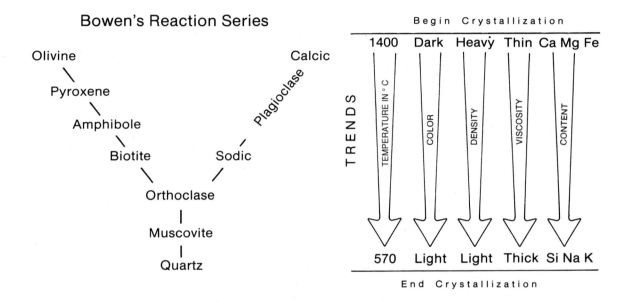

The reaction series shows how rock-forming minerals are related, but no one magma can produce all eight rock-forming minerals as it cools. Different kinds of igneous rocks crystallize from the three kinds of magma: mafic (top of Bowen's Reaction Series), intermediate (middle of Bowen's Reaction Series), and felsic (bottom of Bowen's Reaction Series). The figure on the next page shows how the three kinds of magma are related to Bowen's Reaction Series. If you learn to sketch this out, or have it clearly in your mind's eye, then igneous rock identification is much easier because then you have a systematic way of relating color, composition, and cooling history, as in the following diagram, to get the rock name.

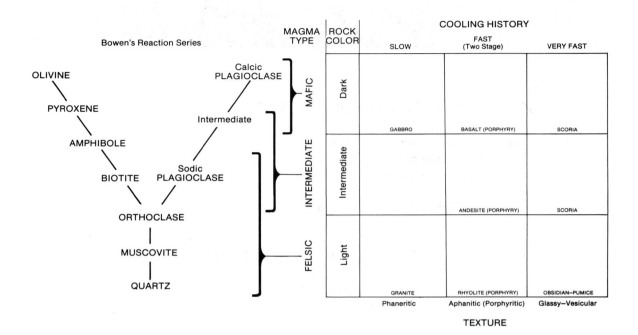

TEXTURE

Igneous Rocks

THIS LABORATORY ASSUMES YOU KNOW OR CAN DO THE FOLLOWING:

1. Define the different igneous rock textures—**glassy, crystalline, vesicular,** and **pyroclastic; phaneritic, aphanitic,** and **porphyritic;** and **groundmass** and **phenocryst**—and describe how these relate to cooling history.

2. Describe the color and mineral composition of mafic, intermediate, and felsic magmas.

3. Describe why an igneous rock classification based on TEXTURE and COLOR is not as good as one based on TEXTURE and COMPOSITION.

4. Sketch Bowen's Reaction Series and describe the trends in temperature, color, density, viscosity, and composition within it.

If you do not know the answers to these find out now. Read the Preliminary to Igneous Rocks, your notes, your textbook, or ask your neighbor.

ORGANIZATION

There are three parts to this laboratory, beginning with two levels of igneous rock classification and identification and ending with some interpretations of igneous rocks. A summary of the subjects in each part is as follows:

PART ONE—IGNEOUS ROCK CLASSIFICATION   Two levels of classification are presented here. *Level One*, based on color and texture (key on page 31), is fast and simple, and names the more obvious phaneritic, aphanitic, porphyritic, glassy, and vesicular igneous rocks. A section on making decision trees for difficult specimens is included. The classification here can be used by itself and will lead directly to the interpretations in Parts Two and Three.

Level Two classification is a more precise classification based on composition and texture in phaneritic rocks (key on page 35). This level can be used for phaneritic rocks without doing Level One but does not include rocks of any other textures. This level also has a section on Bowen's Reaction Series and a section on making decision trees to identify difficult specimens.

PART TWO—INTERPRETATIONS OF EARTH PROCESSES FROM IGNEOUS ROCKS
Using the foldout chart "The Distribution of Igneous Rocks" in the pocket at the back of the manual, this explores how and why igneous rocks are found in different parts of the earth. Can be done following either level of classification.

PART THREE—CRITICAL REASONING PROBLEMS   Questions designed to help you learn to know and understand igneous rocks well enough to do well on a test. Some problems can be done directly following Part One. Other problems require having completed Part Two.

┌─*Part One*───

IGNEOUS ROCK
CLASSIFICATION

└──

LEVEL ONE: CLASSIFICATION ON TEXTURE AND COLOR

At this level you will do two things. First, divide your rocks into groups based on texture and color. Second, identify the major kinds of igneous rocks.

☐   (1)   Get a tray of igneous rocks from your instructor and the foldout chart "Igneous Rock Texture and Color" from the pocket at the back of the manual. Work together in pairs.

☐   (2)   SEPARATE ROCKS BY TEXTURE: Divide your rocks into three piles: first pile, phaneritic rocks with crystals you can see with your eye; second pile, aphanitic rocks so fine grained you cannot see crystals; third pile, glassy, vesicular, and pyroclastic rocks.

> **OBSERVE:** The porphyritic rocks should for right now be placed in the pile with the aphanitic rocks. This is because the name of a porphyry is the name of the aphanitic portion of the rock with the term "porphyry" added at the end.

☐   (3)   Ask your instructor to check your piles based on texture.

☐   (4)   ARRANGE YOUR ROCKS ON THE FOLDOUT CHART BY TEXTURE AND COLOR.

STEP A   Begin with *phaneritic* rocks and arrange them on the foldout chart by color, light on the left and dark on the right.

> **OBSERVE:** The intermediate phaneritic rocks (diorites) may look like dark-colored rocks at first. They have a "pepper and salt" appearance; that is, a mixture of mafic minerals and light-colored plagioclase. To the eye, however, the mafics appear dominant, and the rock appears dark even though it is intermediate in composition. Notice the 50% abundance circle in the chart on page 34; it looks dark, but is 50% light. Mafic igneous rocks have dark plagioclase, making the whole rock dark.

STEP B   Arrange the *aphanitic* rocks on the foldout chart by color, light on the left and dark on the right.

OBSERVE: There are a few aphanitic rocks for which color will just not work for classification, and if you try to use color, the identification will be wrong. There is no way to know this ahead of time. You will just have to memorize these specimens. They include:

Obsidian: Glassy, and black or red. It belongs in the light-colored felsic category because its chemistry is like those rocks. Obsidian is dark because it is a glass with many impurities which absorb the light, making it dark.

Peridotite: In Level One classification peridotite includes igneous rocks with all mafic minerals and no feldspars. Some types are mostly olivine (the rock **dunite**) and appear pale green. These rocks are ultramafic and cannot be placed with the light-colored rocks.

Scoria: Ranges from dark red to black, but its composition ranges from intermediate to mafic.

OBSERVE: Rhyolite and andesite *cannot* be precisely identified in hand specimen; their color is just too variable and overlapping. The safest course is to step back to a less precise identification. The term *felsite* is a catchall that includes both rhyolites and andesites.

STEP C Arrange the glassy, vesicular, and pyroclastic rocks on the foldout chart by color, light on the left and dark on the right.

OBSERVE: Some pyroclastic rocks require experience to recognize and identify. Make the best arrangement you can, but then ask for help.

OBSERVE: There is a complete range of vesicular rocks from those that have only a few vesicles, are heavy, and belong to the aphanitic rocks to rocks with many vesicles, and they are all properly classified as vesicular. Which category a specimen belongs in is a matter of judgement. Classify them as best you can, but then ask for help.

☐ (5) Ask your instructor to check your arrangement of rocks on the chart.

☐ (6) IDENTIFY INDIVIDUAL ROCKS. Use the "Key to the Identification of Igneous Rocks Based on Color and Texture," page 31, to name the rocks on your chart. You can make a quick reference to the rocks you have studied by writing their names in the boxes in the middle of the Igneous Rock Mineral Composition Diagram on page 32. Note that the boxes for felsic rocks are not identical to those on the foldout chart; just write the names in the center of the felsite spaces. Ignore the diagrams above and below the boxes.

☐ (7) Make sure your rock names are in the right places. Compare with your neighbor.

DECISION TREES FOR IDENTIFYING IGNEOUS ROCKS

In the laboratory on minerals we contrasted the brute force method for learning minerals and rocks ("I'll just stare at the specimens so long maybe I'll recognize them on the exam") with the hypothesis-test method (see pages 13-14). The hypothesis-test method helps you to understand why rocks are different from each other and to develop a strategy for confidently identifying unknown specimens.

☐   (1)   If they are available, pick out and lay in front of you the following sets of igneous rocks.
   A.   Pumice, obsidian, scoria, vesicular basalt
   B.   Peridotite, gabbro, diorite
   C.   Basalt, obsidian, peridotite
   D.   Granite, rhyolite porphyry, diorite

☐   (2)   On a separate piece of paper draw a decision tree like those you did in the minerals lab, pages 16-17.
   When finished, compare your decision trees with ours at the back of the chapter. Please do not look at our decision trees before you draw your own. If you are unable to draw a decision tree that works, then you should get help.

☐   (3)   Now pick out any other igneous rocks which confuse you. Ask, "What distinctive characteristics will separate these rocks?" Develop a hypothesis-test strategy which *you* can use to separate and identify the specimens.

LEVEL TWO: CLASSIFICATION BY COMPOSITION AND TEXTURE

In Level One classification, igneous rocks were identified by texture and color. Color is a first approximation for identifying igneous rocks but will eventually let you down. For example, both rhyolites and andesites can be light gray, and some diorites can look very close to a granite while others look close to a gabbro. Plus, as you now know, color in minerals is very variable and not always useful for identification. It is no different with igneous rocks.

Besides, what a geologist really wants to know about an igneous rock is its chemistry and only composition can tell us that. A better classification must include both texture and composition, but then we must be able to see and identify the minerals. Fine-grained rocks can only be studied in thin section (paper-thin slices of rock looked at through a specialized polarizing microscope), so we are limited to coarse-grained rocks here.

LEVEL TWO IDENTIFICATION

(1)   Get a tray of igneous rocks from your instructor. Work together in pairs. Remove all the phaneritic (coarse-grained) rocks for study.

(2)   Use the data sheets on pages 36-38 to record your observations for each rock.

(3)   IDENTIFY ALL THE MINERALS you can find in the rock using a hand lens or lower-power microscope and list them on the data sheet. Take the time to look at the rock thoroughly and carefully. It is very important that you be able to distinguish between the pairs of confusing minerals, such as

   A.   Pyroxene and amphibole   C.   Amphibole and biotite
   B.   Plagioclase and orthoclase   D.   Olivine and quartz

The classification depends on accurately identifying these minerals.

KEY TO LEVEL ONE CLASSIFICATION
Identifying Igneous Rocks by Texture and Color

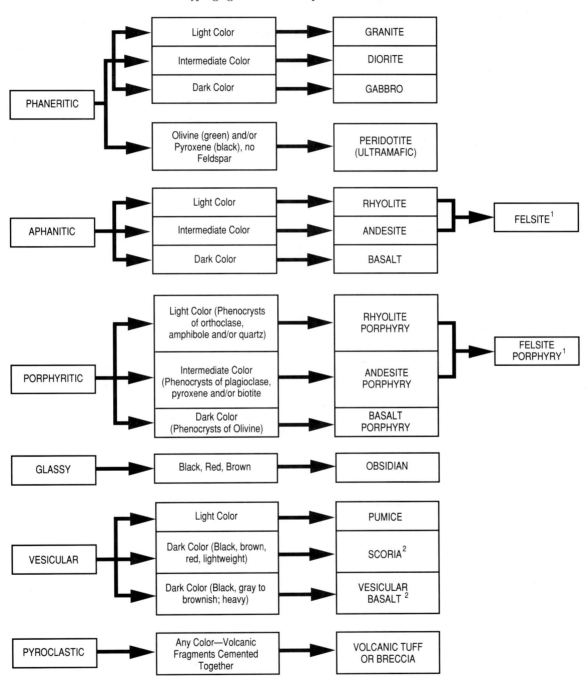

[1] Felsite is a general rock name that includes both the igneous rocks rhyolite and andesite. The term *felsite* is used when it is impossible to distinguish between these two.

[2] There exists a complete gradation between rocks that can be classified as scoria and those classified as vesicular basalt. Which it is is a matter of judgment, but do your best.

IGNEOUS ROCK MINERAL COMPOSITION DIAGRAM

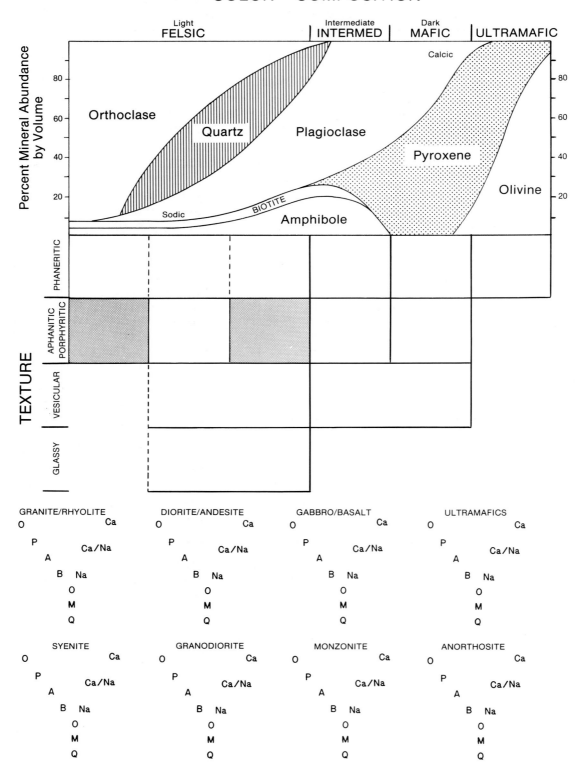

> **OBSERVE:** If separating these minerals in the rock is giving you fits try two things. First, get good samples of these minerals from the minerals tray and look at them under the microscope until you can recognize fragments in the rock. Second, make the best identification you can, and then ask your instructor to check it. If you identified the mineral incorrectly, find out what it is and study it until you learn what the mineral looks like in a rock.

(4) ESTIMATE THE PERCENT ABUNDANCE OF EACH MINERAL IDENTI-FIED IN STEP 3. You need to learn to estimate percent abudance of minerals in a rock. On page 34 is a percentage composition chart. The circles imitate the view seen through a microscope. The black marks represent one kind of mineral, and the percent number under the circle is the percent of the area of the circle taken up by the black marks.

Observing the rock under a microscope or hand lens look for one mineral and compare the area it takes up with the different percent abundance charts until you have what you judge to be the right percent. Write that number down on the data sheets on pages 36–38. Do the same thing for the next mineral, and the next, until all have been estimated. Your percentages should add up to 100% plus or minus about 5%.

> **OBSERVE:** Estimating percentage abundance is a learned skill. Techniques do exist for making estimates with minimal error, but just using a simple microscope there are no "right" answers—just estimates, some of which are better than others. The best strategy is to work with a partner and each of you make an estimate independently. Then compare; if you disagree by some large percent explain to each other how you made your estimate and see if you can come to an estimate acceptable to both. Check your estimates with other class members and your instructor.

(5) IDENTIFY PHANERITIC IGNEOUS ROCKS BY COMPOSITION. There are two ways you can identify an igneous rock once you know the mineral abundances. The first way is to use the *key* on page 35. This key breaks the continuous variation in mineral composition into arbitrary categories.

The second way of classifying igneous rocks is to use the Igneous Rock Mineral Composition Diagram at the top of page 32. Because this diagram is visual, it is sometimes faster and easier to use than a key. Note that a vertical line drawn anywhere from bottom to top of the chart adds up to 100% and the percent abundance of each mineral can be quickly and easily scaled off.

> **OBSERVE:** Three rocks—monzonite, anorthosite, and pyroxenite—cannot be identified using the diagram on page 32. They just have to be memorized. The failure of this chart to consistently classify all the rocks indicates the system is not perfect. Better classifications use ternary diagrams, but these are introduced only for sedimentary rock classification.

(6) Ask your instructor to check your identifications.

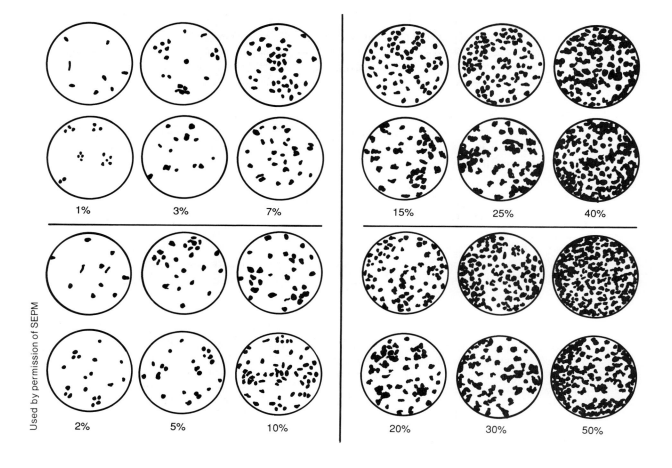

KEY TO LEVEL TWO CLASSIFICATION
Identifying Phaneritic Igneous Rocks By Composition

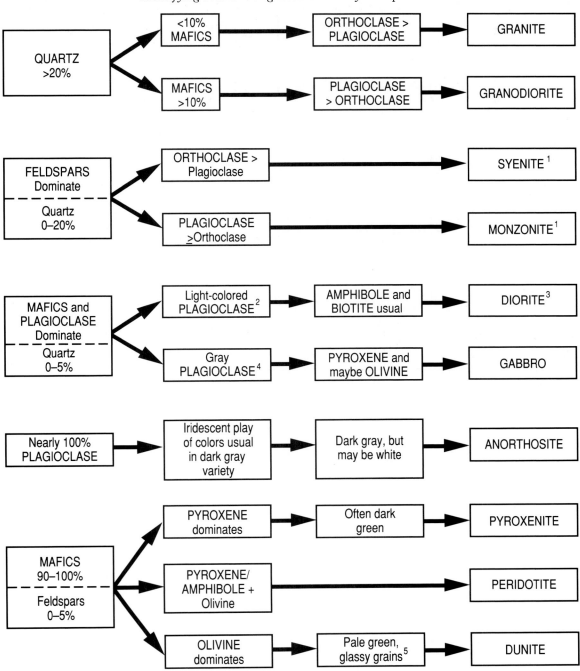

[1] If quartz is present between 5–20% call the rock a quartz syenite or quartz monzonite.
[2] Plagioclase is usually light-colored, but other colors are possible.
[3] Quartz sometimes between 5–20% in diorites; call them quartz diorites.
[4] This plagioclase is often gray, but other varieties are possible.
[5] Almost always has small black crystals of chromite.

LEVEL TWO DATA SHEETS

MAJOR MINERALS	MAFICS (list in % order)	PLAGIOCLASE COLOR	CIRCLE MINERALS
Quartz % _____	_____ ____ %		O Ca
Mafics % _____	_____ ____ %		P Ca/Na
Plagioclase % _____		ROCK NAME	A Na
Orthoclase % _____	_____ ____ %		B
Total %	Total ____ %		O M Q

MAJOR MINERALS	MAFICS (list in % order)	PLAGIOCLASE COLOR	CIRCLE MINERALS
Quartz % _____	_____ ____ %		O Ca
Mafics % _____			P Ca/Na
Plagioclase % _____	_____ ____ %	ROCK NAME	A Na
Orthoclase % _____	_____ ____ %		B
Total %	Total ____ %		O M Q

MAJOR MINERALS	MAFICS (list in % order)	PLAGIOCLASE COLOR	CIRCLE MINERALS
Quartz % _____	_____ ____ %		O Ca
Mafics % _____			P Ca/Na
Plagioclase % _____	_____ ____ %	ROCK NAME	A Na
Orthoclase % _____	_____ ____ %		B
Total %	Total ____ %		O M Q

MAJOR MINERALS	MAFICS (list in % order)	PLAGIOCLASE COLOR	CIRCLE MINERALS
Quartz % _____	_____ ____ %		O Ca
Mafics % _____			P Ca/Na
Plagioclase % _____	_____ ____ %	ROCK NAME	A Na
Orthoclase % _____	_____ ____ %		B
Total %	Total ____ %		O M Q

LEVEL TWO DATA SHEETS

MAJOR MINERALS	MAFICS (list in % order)	PLAGIOCLASE COLOR	CIRCLE MINERALS
Quartz % _____ Mafics % _____ Plagioclase % _____ Orthoclase % _____ Total %	_____ _____ % _____ _____ % _____ _____ % Total _____ %	ROCK NAME	O Ca P Ca/Na A Na B O M Q

MAJOR MINERALS	MAFICS (list in % order)	PLAGIOCLASE COLOR	CIRCLE MINERALS
Quartz % _____ Mafics % _____ Plagioclase % _____ Orthoclase % _____ Total %	_____ _____ % _____ _____ % _____ _____ % Total _____ %	ROCK NAME	O Ca P Ca/Na A Na B O M Q

MAJOR MINERALS	MAFICS (list in % order)	PLAGIOCLASE COLOR	CIRCLE MINERALS
Quartz % _____ Mafics % _____ Plagioclase % _____ Orthoclase % _____ Total %	_____ _____ % _____ _____ % _____ _____ % Total _____ %	ROCK NAME	O Ca P Ca/Na A Na B O M Q

MAJOR MINERALS	MAFICS (list in % order)	PLAGIOCLASE COLOR	CIRCLE MINERALS
Quartz % _____ Mafics % _____ Plagioclase % _____ Orthoclase % _____ Total %	_____ _____ % _____ _____ % _____ _____ % Total _____ %	ROCK NAME	O Ca P Ca/Na A Na B O M Q

LEVEL TWO DATA SHEETS

MAJOR MINERALS	MAFICS (list in % order)	PLAGIOCLASE COLOR	CIRCLE MINERALS
Quartz % _____	_____ ___ %		O Ca
Mafics % _____	_____ ___ %		P Ca/Na
Plagioclase % _____	_____ ___ %	ROCK NAME	A
Orthoclase % _____	_____ ___ %		B Na
Total %	Total ___ %		O
			M
			Q

MAJOR MINERALS	MAFICS (list in % order)	PLAGIOCLASE COLOR	CIRCLE MINERALS
Quartz % _____	_____ ___ %		O Ca
Mafics % _____	_____ ___ %		P Ca/Na
Plagioclase % _____	_____ ___ %	ROCK NAME	A
Orthoclase % _____	_____ ___ %		B Na
Total %	Total ___ %		O
			M
			Q

MAJOR MINERALS	MAFICS (list in % order)	PLAGIOCLASE COLOR	CIRCLE MINERALS
Quartz % _____	_____ ___ %		O Ca
Mafics % _____	_____ ___ %		P Ca/Na
Plagioclase % _____	_____ ___ %	ROCK NAME	A
Orthoclase % _____	_____ ___ %		B Na
Total %	Total ___ %		O
			M
			Q

MAJOR MINERALS	MAFICS (list in % order)	PLAGIOCLASE COLOR	CIRCLE MINERALS
Quartz % _____	_____ ___ %		O Ca
Mafics % _____	_____ ___ %		P Ca/Na
Plagioclase % _____	_____ ___ %	ROCK NAME	A
Orthoclase % _____	_____ ___ %		B Na
Total %	Total ___ %		O
			M
			Q

IGNEOUS ROCKS AND BOWEN'S REACTION SERIES

The minerals which compose the different igneous rocks are clearly not the same, but they are not random or haphazard either. All the igneous rock–forming minerals are systematically related to each other in many different ways, including temperature, composition, and color. All these relationships are summarized in Bowen's Reaction Series.

Bowen's Reaction Series is such an important and powerful model for understanding igneous rocks that it is important that you develop a deliberate connection between the relationships expressed in the reaction series and your visual and tactile experience with the minerals and the rocks. The exercises that follow will help you to make these connections. What is important is deliberately explaining to yourself, your partner, or your instructor what the relationships are, and why, as you go through these exercises. The understanding must become intuitive.

☐ (1) Get the foldout diagram of Bowen's Reaction Series from the pocket at the back of the manual, and the eight rock forming minerals from the minerals tray. Place the minerals on the reaction series in their appropriate location.

Carefully and deliberately compare and contrast all the minerals while reciting how the minerals in the reaction series are related by (A) color, (B) crystallization temperature, (C) density, and (D) type of occurrence. These relationships are discussed in the Preliminary, pages 24-26.

☐ (2) At the bottom of the Igneous Rock Mineral Composition Diagram on page 32 are outline drawings of Bowen's Reaction Series. Above each reaction series is the name of an igneous rock group.

Get a sample of each igneous rock listed above a reaction series. One by one examine them under a microscope or with a hand lens. Identify the minerals present and circle the minerals you find on Bowen's Reactions Series. For each mineral estimate its percent abundance using the chart on page 34 and write the number beside the mineral in the reaction series. Make sure your percentages add up to 100% ± 5%.

☐ (3) Have your instructor check your mineral identifications and percent mineral abundances.

DECISION TREES FOR IDENTIFYING IGNEOUS ROCKS

☐ (1) If they are available, pick out and lay in front of you the following sets of igneous rocks.

    A. Granite, diorite, gabbro
    B. Granite, syenite, monzonite
    C. Peridotite, pyroxenite, dunite, anorthosite
    D. Anorthosite, gabbro, peridotite

☐ (2) On a separate piece of paper draw a decision tree like those you did in the mineral lab, pages 16-17. Decision trees should make use of mineral content. **You may not use color.**

When finished compare your decision trees with your instructor's. If you are unable to draw a decision tree which works then you should get help.

☐ (3) Now pick out any other igneous rocks which confuse you. Ask, "What distinctive characteristics will separate these rocks?" Develop a hypothesis-test strategy which **you** can use to separate and identify the specimens.

OBSERVE: The minerals you identified in the minerals lab are frequently just small pieces of rock. You should not arbitrarily place mineral and rock specimens in these different categories. Go back and examine your mineral specimens to see if any can be classified as a rock as well as a mineral.

Part Two

INTERPRETATIONS OF EARTH PROCESSES FROM IGNEOUS ROCKS

One of the important things to understand about igneous rocks is that they are not just found anywhere on the earth. Igneous rocks are always found in particular places for particular reasons. Part of understanding geology is understanding why particular rocks are found where they are.

☐ (1) This section makes more sense if you understand the principles of Bowen's Reaction Series discussed in the Preliminary, pages 24-26.

☐ (2) There are two things you should learn here:
A. To develop a picture, in your mind's eye, of the earth's structure,
B. To be able to accurately place on that picture of the earth's structure, in your mind's eye, various kinds of igneous rocks, and explain why they must be found in that particular place.

When you are done you should be able to point to any part of the cross section, identify it, and say what kind of igneous rock would be found there.

☐ (3) The earth is subdivided by several different criteria, including physical behavior, composition, and tectonics. These subdivisions are briefly described next and you need to be familiar with them. You may wonder how we know about these divisions, most of which are far below the earth's surface. A large body of theory lies behind each of them, and your lecture instructor and textbook will explore that. Here we are interested only in describing them enough so you can find them on a cross section.

SUBDIVISIONS OF THE EARTH

Earth Structure Based on Physical Properties and Composition In cross section the earth is divided into a number of layers. At the center of the earth is the core. It is made largely of iron and nickel and is molten. We will not discuss the core further.

The rest of the earth outside the core is divided in two ways, one on physical behavior and the other on composition. This is sometimes confusing because these two methods of division have different terms applied to them and the divisions overlap. Based on *physical behavior* the area outside the core is divided into an **asthenosphere** and a **lithosphere.** The asthenosphere is on the inside and is hot, plastic rock which flows very slowly (centimeters per year) in convection cells. The asthenosphere has many subdivisions, and its exact structure and the way the convection cells work are not clearly understood yet. The lithosphere is on the outside and is cold, brittle rock which breaks when stressed.

Based on *composition* the area outside the core is divided into a **mantle** and a **crust.** The uppermost part of the mantle is included in the lithosphere because it is (relatively) cool and brittle in behavior. The deeper parts of the mantle belong to the asthenosphere because, there, it is hot and behaves plastically. The subdivisions of the earth based on physical behavior and composition are illustrated in the drawing below.

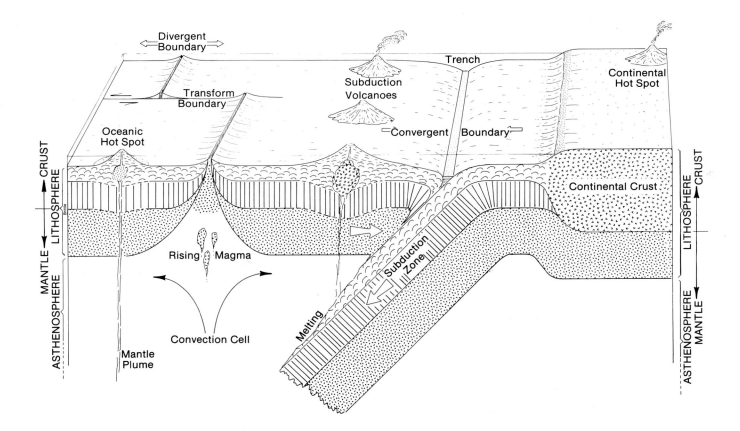

Tectonic Divisions of the Lithosphere   **Tectonics** is concerned with deformation in the earth, the forces which produce the deformation and the pieces of the earth which cause or result from the deformation. **Plate tectonics** is the theory that the earth's lithosphere is divided into "plates," or pieces, which behave as tectonic units. Today the lithosphere is composed of several dozen plates which float on the mantle, like slabs of ice on a pond (see the figure on page 41, which is a small version of chart "The Distribution of Igneous Rocks" in pocket at back of manual).

In plate tectonic theory, earth history at its simplest is one of plates rifting into pieces and diverging apart and new ocean basins being born, followed by motion reversal, convergence back together, plate collision, and mountain building. Most geologic activity occurs at the boundaries between plates.

There are three kinds of plate boundaries: (1) **divergent boundaries** where plates are moving apart and new crust is being created, (2) **convergent boundaries** where plates are moving together and crust is subducting, that is, diving down into the mantle, and (3) **transform boundaries** where plates slide past one another. One additional kind of geologic activity occurs within plates; these are **hot spots.** A hot spot is a plume of hot magma that rises toward the earth's surface from within the mantle. Hot spots are randomly scattered over the earth's surface.

With this brief background let us look at the distribution of igneous rocks on the earth.

☐   (4)   Get a tray of igneous rocks and the foldout chart in the pocket at the back of the manual titled "The Distribution of Igneous Rocks."

☐   (5)   As you read through the descriptions that follow, place an appropriate igneous rock from your tray on the chart over the place it belongs. We will not examine all the igneous rocks—just the more important or abundant ones.

The crust of the earth is made of two parts, **continents** and **ocean basins**. The continents are composed of granitic rocks.

Name:	**GRANITE**—a felsic igneous rock
Texture:	Phaneritic
Color:	Light or white to pink or red (depending on feldspar)
Composition:	Quartz, feldspars (orthoclase and/or sodic plagioclase), biotite, muscovite, and/or amphibole (low on Bowen's Reaction Series; low in density).
Interpretation:	Large crystals—slow cooling deep underground in batholiths. Forms the bulk of the continents.

Examine the specimen with a handlens or microscope. You should see large crystals of quartz and feldspar, with possible smaller amounts of biotite, amphibole, or muscovite. Can you identify the feldspar? Is it plagioclase or orthoclase? Check your determinations with your instructor.

Granite is probably the most familiar of all igneous rocks. When polished it is beautiful and is commonly used as a building stone. But granite is important because it forms the bulk of the continents. The minerals in granite are of lower density than are minerals in mafic rocks because they come from the bottom of Bowen's Reaction Series. Thus, granite "floats" high in the mantle, just as an empty boat floats high in the water. When nothing is artificially holding a piece of continent up, as occurs in mountain building, a continent floats just a few hundred feet above sea level. The origin of these floating continental blocks is a major problem in geology since, to date, they have not been detected on any other planet in the solar system.

☐   (6)   Place a piece of granite over continental crust on your cross section.

The **ocean basins** are composed of the mafic igneous rocks, basalt near the surface and gabbro at depth.

Name: **BASALT** or **BASALT PORPHYRY**—a mafic igneous rock
Texture: Aphanitic or aphanitic porphyritic
Color: Dark
Composition: Pyroxene, calcic plagioclase, and sometimes olivine (high in Bowen's Reaction Series; high in density).
Interpretation: Small crystals—rapid cooling at or near the earth's surface. Composes the ocean floors and makes oceanic volcanoes (like Hawaii), but does occur locally on the continents from hot spot volcanoes.

Basalt is typically dark and nondescript in appearance since all the minerals are too small to be seen by eye. Can you see any crystals? Try a microscope or hand lens. In the early days of geology there was great difficulty interpreting basalt; one theory was that it precipitated out of seawater. Today we know the small grain size is the result of rapid cooling and solidification of magma at or near the earth's surface. If phenocrysts are present, they indicate two cooling stages. Magma under pressure contains dissolved gas much like a carbonated beverage. When magma is extruded, the pressure is released and the bubbles begin to escape. Sometimes this leaves behind holes called **vesicles**. When the lava erupts under water the quick chilling causes the lava to form "pillows," bulbous-shaped blobs of lava.

The dark color results from high concentrations of mafic minerals near the top of Bowen's Reaction Series. Basalt is thus of higher density than granitic rocks and "floats" low in the mantle, in fact, when it is not being artificially held up or down, about $5\frac{1}{2}$ kilometers below sea level.

Basalt most commonly forms at divergent plate boundaries when convection cells bring magma up from the mantle.

☐   (7)   Place a piece of basalt over the upper portion of the oceanic crust on your cross section.

Name: **GABBRO**—a mafic igneous rock
Texture: Phaneritic
Color: Dark
Composition: Pyroxene, calcic plagioclase, and sometimes olivine (high in Bowen's Reaction Series; high in density).
Interpretation: Large crystals—slow cooling deep underground. Composes the deep crustal rocks under the basalt floor of the ocean, but also found in large isolated intrusive bodies on the continents associated with hot spots.

You may have heard of the building stone called "black granite." Clearly from what you now know of igneous rocks that is not possible. What is called "black granite" is usually gabbro. Gabbro has the same composition as basalt but just cools slowly enough for the crystals to grow large.

☐   (8)   Place a piece of gabbro below the basalt on your cross section over the lower portion of the oceanic crust. Do not place it over the mantle.

Although basalt and gabbro are most common in the ocean basins they are also found on parts of the continents. This is the result of hot spot activity, that is, plumes of magma rising from the mantle and punching a hole through the earth's crust. The mafic magma may form volcanoes or volcanic fields or stacks of lava flows thousands of feet thick (plateau or flood basalts).[1] Most of the isolated volcanic islands within the ocean basins are also the result of hot spots, such as the islands of Hawaii.[2]

☐   (9)   Place a piece of mafic igneous rock on your cross section over one of the hot spots. If you do not have a spare piece take the piece of basalt from the ocean crust and put it there.

Name:	**PERIDOTITE** or **DUNITE**—ultramafic igneous rocks
Texture:	Phaneritic
Color:	May appear light if dunite (90%+ olivine) or mixed light/dark if olivine and pyroxene (peridotite).
Composition:	Pyroxene and/or olivine (upper left arm of Bowen's Reaction Series, very high density).
Interpretation:	Large crystals—slow cooling deep underground. Composes the very deepest portions of the ocean basins (under the basalt and gabbro) and the earth's mantle. Found on continents only where vigorous mountain building thrusts pieces of mantle onto a continental edge.

What makes an ultramafic rock so unusual among igneous rocks is that it has no feldspar. Look at the Igneous Rock Mineral Composition Chart on page 32 and you can see that feldspars of one kind or another are important in all other igneous rocks. Look carefully at your specimen with a hand lens or microscope and try to find some feldspar—there should be none. Look for the olivine with a hand lens or microscope; do not confuse it with quartz, a common mistake at first.

Peridotite and dunite come from the very top left of Bowen's Reaction Series and have the highest density of the eight rock-forming minerals. For this reason they sink or lie deepest within the earth and compose much of the mantle.[3] It is on these rocks that the continental granite and oceanic basalt/gabbro float.

---

[1] Some common examples of hot-spot volcanics are Craters of the Moon in Idaho, the lava flows along the Snake River plain, the volcanics and geysers at Yellowstone Park (all caused by the same hot spot which has migrated eastward through time), Ship Rock in the Four Corners region of the southwest, and the Devils Tower in Wyoming. In the East all the hot-spot volcanics are ancient but include basalt flows in the Blue Ridge mountains of Virginia.

[2] Some volcanic islands like Hawaii are known for the beaches of black sand composed of particles of basalt. Other volcanic islands have beaches made of white limestone sand; these are usually old extinct volcanic islands which have sunk below sea level. The white limestone sand is from the reefs which have grown up around the volcano and now are the only thing remaining at the surface.

[3] At deeper layers within the mantle the pressures become so great that peridotite or dunite are converted into other high-pressure minerals and rocks, such as eclogite studied in the metamorphic lab. Your instructor may be able to show you an example of such a rock.

☐ (10) Place a piece of peridotite and/or dunite on your cross section over the mantle.

Name:	**ANDESITE** or **ANDESITE PORPHYRY**
Texture:	Aphanitic or aphanitic porphyritic
Color:	Intermediate
Composition:	Intermediate plagioclase, biotite, amphibole, pyroxene (middle of Bowen's Reaction Series, intermediate density).
Interpretation:	Small crystals—rapid cooling at or near the earth's surface. Forms only in volcanoes above subducting oceanic crust; may be within an ocean basin (e.g., Japan, Aleutian islands) or along a continent edge (e.g., Cascade volcanoes and Andes).

Name:	**DIORITE**
Texture:	Phaneritic
Color:	Intermediate—mixture of dark and light crystals
Composition:	Intermediate plagioclase, amphibole, some pyroxene, trace quartz (middle of Bowen's Reaction Series, intermediate density).
Interpretation:	Large crystals—slow cooling deep underground. Same locations as andesite.

Andesite (porphyry) and diorite are both intermediate igneous rocks with a composition in the middle of Bowen's Reaction Series. The first just cools quickly, and the second slowly. These rocks form under only one set of geological conditions, a convergent plate boundary. At a convergent boundary two plates are coming together and something has to give. What gives is the edge of one plate which sinks, or subducts, into the mantle. If the edge of both plates are composed of oceanic lithosphere, it is probably a matter of chance which plate edge subducts, but if one plate edge is ocean basalt/gabbro lithosphere and one is a granite continent, it will always be the oceanic lithosphere which subducts. The granite continent is just too low in density (low in Bowen's Reaction Series) to go down.

The lithosphere plate which is subducting into the mantle heats up because of friction and other processes. In time it is hot enough and begins to melt and an intermediate magma is formed. The magma is lower in density than the surrounding ultramafic mantle rock both because it is hot and because intermediate magma, being in the middle of Bowen's Reaction Series, has lower density than does magma at the top of Bowen's Reaction Series. Thus the intermediate magma rises toward the surface where it ponds and forms a diorite batholith. Some of the magma reaches the surface and forms an andesite (porphyry) volcano. These volcanoes are typically enormous.[1]

☐ (11) Place a piece of diorite over the batholith above the subduction zone on your cross section, and over the volcano above that a piece of andesite (porphyry).

---

[1] Typical examples are Mt. St. Helens, Mt. Shasta, and Mt. Rainier in the Cascade Mountains of Washington, Oregon, and northern California. Others include the volcanoes which make up such islands as Japan, Borneo, Sumatra, New Guinea, and the Aleutian Islands of Alaska.

You have several rocks left over in your tray. The rhyolite (porphyry), obsidian, and pumice all have a composition near the bottom of Bowen's Reaction Series, Just as does granite and, therefore, are found on the continents. Other rocks not discussed are usually closely related to ones discussed and would be closely related tectonically.

☐   (12)   Place all the leftover rocks in your tray on the cross section with the rocks they are most closely associated with.

☐   (13)   When you are done call your instructor and show her or him how you have placed the rocks on the chart. Be prepared to explain why the rocks are found on the earth where they are.

PRACTICING IGNEOUS ROCK IDENTIFICATION AND INTERPRETATION

There is only one way to learn to identify and interpret rocks—practice, practice, practice. Many igneous rocks which are classified the same may look superficially different at first. What you must learn to do is see through the superficial differences to the characteristics which are important in classification and interpretation. So the more rocks you handle and study the better you will become.

☐   (1)   Your instructor will provide you with a new group of igneous rocks, ones you have not yet seen. Carefully examine each igneous rock and record the rock's texture, color, and mineral composition on the charts on pages 47–50.

☐   (2)   Think about the processes by which each igneous rock forms and record the **cooling history** and **location on earth** (use foldout chart "The Distribution of Igneous Rocks") for each rock on the charts on pages 47–50.

☐   (3)   If you are confused or unsure of what you are doing, or how to do it, *ask for help*. Now is the time to be unsure, make mistakes, and learn from them.

☐   (4)   When you are done call your instructor and show her or him how you have placed the rocks on the chart. Be prepared to explain why the rocks are found on the earth where they are.

DATA SHEETS FOR PRACTICING
IGNEOUS ROCK CLASSIFICATION AND INTERPRETATION

CHARACTERISTICS   SPECIMEN NUMBER   IGNEOUS ROCK NAME

TEXTURE	COLOR	MINERAL COMPOSITION
COOLING HISTORY		
LOCATION(S) ON EARTH		

TEXTURE	COLOR	MINERAL COMPOSITION
COOLING HISTORY		
LOCATION(S) ON EARTH		

TEXTURE	COLOR	MINERAL COMPOSITION
COOLING HISTORY		
LOCATION(S) ON EARTH		

TEXTURE	COLOR	MINERAL COMPOSITION
COOLING HISTORY		
LOCATION(S) ON EARTH		

TEXTURE	COLOR	MINERAL COMPOSITION
COOLING HISTORY		
LOCATION(S) ON EARTH		

DATA SHEETS FOR PRACTICING
IGNEOUS ROCK CLASSIFICATION AND INTERPRETATION

CHARACTERISTICS   SPECIMEN NUMBER   IGNEOUS ROCK NAME

TEXTURE	COLOR	MINERAL COMPOSITION
COOLING HISTORY		
LOCATION(S) ON EARTH		

TEXTURE	COLOR	MINERAL COMPOSITION
COOLING HISTORY		
LOCATION(S) ON EARTH		

TEXTURE	COLOR	MINERAL COMPOSITION
COOLING HISTORY		
LOCATION(S) ON EARTH		

TEXTURE	COLOR	MINERAL COMPOSITION
COOLING HISTORY		
LOCATION(S) ON EARTH		

TEXTURE	COLOR	MINERAL COMPOSITION
COOLING HISTORY		
LOCATION(S) ON EARTH		

**DATA SHEETS FOR PRACTICING
IGNEOUS ROCK CLASSIFICATION AND INTERPRETATION**

CHARACTERISTICS

SPECIMEN
NUMBER

IGNEOUS
ROCK
NAME

TEXTURE	COLOR	MINERAL COMPOSITION
COOLING HISTORY		
LOCATION(S) ON EARTH		

TEXTURE	COLOR	MINERAL COMPOSITION
COOLING HISTORY		
LOCATION(S) ON EARTH		

TEXTURE	COLOR	MINERAL COMPOSITION
COOLING HISTORY		
LOCATION(S) ON EARTH		

TEXTURE	COLOR	MINERAL COMPOSITION
COOLING HISTORY		
LOCATION(S) ON EARTH		

TEXTURE	COLOR	MINERAL COMPOSITION
COOLING HISTORY		
LOCATION(S) ON EARTH		

**DATA SHEETS FOR PRACTICING
IGNEOUS ROCK CLASSIFICATION AND INTERPRETATION**

CHARACTERISTICS	SPECIMEN NUMBER	IGNEOUS ROCK NAME

Block 1

TEXTURE	COLOR	MINERAL COMPOSITION
COOLING HISTORY		
LOCATION(S) ON EARTH		

Block 2

TEXTURE	COLOR	MINERAL COMPOSITION
COOLING HISTORY		
LOCATION(S) ON EARTH		

Block 3

TEXTURE	COLOR	MINERAL COMPOSITION
COOLING HISTORY		
LOCATION(S) ON EARTH		

Block 4

TEXTURE	COLOR	MINERAL COMPOSITION
COOLING HISTORY		
LOCATION(S) ON EARTH		

Block 5

TEXTURE	COLOR	MINERAL COMPOSITION
COOLING HISTORY		
LOCATION(S) ON EARTH		

DATA SHEETS FOR PRACTICING
IGNEOUS ROCK CLASSIFICATION AND INTERPRETATION

CHARACTERISTICS   SPECIMEN   IGNEOUS
                            NUMBER   ROCK
                                                  NAME

TEXTURE	COLOR	MINERAL COMPOSITION
COOLING HISTORY		
LOCATION(S) ON EARTH		

TEXTURE	COLOR	MINERAL COMPOSITION
COOLING HISTORY		
LOCATION(S) ON EARTH		

TEXTURE	COLOR	MINERAL COMPOSITION
COOLING HISTORY		
LOCATION(S) ON EARTH		

TEXTURE	COLOR	MINERAL COMPOSITION
COOLING HISTORY		
LOCATION(S) ON EARTH		

TEXTURE	COLOR	MINERAL COMPOSITION
COOLING HISTORY		
LOCATION(S) ON EARTH		

⌐Part Three─────────────────────────────

CRITICAL REASONING
PROBLEMS───────

In science it is not good enough to just *feel* that you know the right answer. You must also be able to explain factually and logically why one answer is right, and why another answer is wrong, and do it in a way that would convince a skeptic. The questions here are designed to help you learn to know and understand igneous rocks well enough that you can explain what you know to someone else (which is, of course, what a test is all about).

☐   (1)   On a separate piece of paper write answers to the following critical reasoning problems about igneous rocks following these instructions:

          A.   For each problem there is a statement followed by several possible answers. For each possible answer you must indicate whether you REJECT it or ACCEPT it.

          B.   Logically and factually, in *writing*, explain your acceptance or rejection of *each* and *every* answer. There is NO CREDIT for a "right" answer, only for the analysis.

          C.   For each problem there are one ACCEPT and two REJECTS. Some answers will differ by subtle distinctions, but you must find the *best* answer and explain why the other answers are less correct. It is more important that you be able to explain why the wrong answers are wrong and the right answers are right.

          D.   You may discuss the problem with classmates, but when you write your analysis, it must be your own thinking, in your own writing.

☐   (2)   For each problem, find examples of each of the rocks and minerals and lay them out so you can examine them while you formulate and write your reasons for acceptance or rejection.

PROBLEMS OF TEXTURE, COMPOSITION, COOLING HISTORY

Problem Number One

ACCEPT one, REJECT two. A rock which forms from slow cooling, composed of the minerals pyroxene, calcic plagioclase, and possibly olivine would be:

_____ 1. Diorite

_____ 2. Gabbro

_____ 3. Granite

Problem Number Two

ACCEPT one, REJECT two. Diorite most likely forms under the following conditions:

_____ 1. Rapid cooling at the earth's surface.

_____ 2. Slow cooling in a batholith.

_____ 3. Two-stage cooling in a volcano.

Problem Number Three

ACCEPT one, REJECT the two statements which are not true for a granite.

_____ 1. Forms on the continents deep underground.

_____ 2. Is rich in iron, silica, and calcium.

_____ 3. Is related to an andesite porphyry.

Problem Number Four

ACCEPT one, REJECT two. One of these sets of rocks is related by composition.

_____ 1. Peridotite, scoria, obsidian

_____ 2. Pumice, obsidian, rhyolite

_____ 3. Diorite, andesite porphyry, basalt

Problem Number Five

ACCEPT one, REJECT two. A rock dominated by sodic plagioclase and amphibole, formed by slow cooling deep underground would be:

_____ 1. Andesite porphyry

_____ 2. Diorite

_____ 3. Gabbro

PROBLEMS WHICH INCLUDE IGNEOUS ROCK DISTRIBUTION

Problem Number Six

ACCEPT one, REJECT the two statements which are *not true* about an andesite porphyry.

_____ 1. Has two-stage cooling, one deep underground, the other at the earth's surface.

_____ 2. Is found in hot spot volcanoes in the ocean basins.

_____ 3. Is high in olivine.

Problem Number Seven

ACCEPT one, REJECT two. Andesite most likely forms under the following conditions:

_____ 1. In a hot spot volcano.

_____ 2. In a subduction volcano.

_____ 3. In a batholith.

Problem Number Eight

ACCEPT one, REJECT two. Below is a block diagram of a portion of the earth's lithosphere with certain locations identified by letter. An andesite porphyry most likely forms under the following conditions.

_____ 1. Location A

_____ 2. Location B

_____ 3. Location C

BLOCK DIAGRAM FOR CRITICAL REASONING

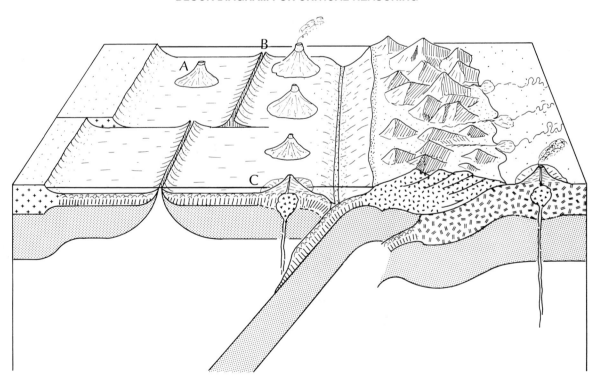

Problem Number Nine

ACCEPT one, REJECT two. On the preceding page is a block diagram of a portion of the earth's lithosphere with certain locations identified by letter. You have the task of collecting basalt. At which locality are you most likely to find the basalt?

_____ 1. Location A

_____ 2. Location B

_____ 3. Location C

PROBLEMS INVOLVING BOWEN'S REACTION SERIES

Problem Number Ten

ACCEPT one, REJECT two. To the right is an outline of Bowen's Reaction Series. A rock composed of the circled minerals would be:

_____ 1. Granite

_____ 2. Gabbro

_____ 3. Andesite

Problem Number Eleven

ACCEPT one, REJECT two. To the right is an outline of Bowen's Reaction Series. A rock composed of the circled minerals would be:

_____ 1. Peridotite

_____ 2. Gabbro

_____ 3. Diorite

Problem Number Twelve

ACCEPT one, REJECT two. To the right is an outline of Bowen's Reaction Series. A rock composed of the circled minerals formed by rapid cooling would be:

_____ 1. Rhyolite

_____ 2. Diorite

_____ 3. Pumice

Problem Number Thirteen

ACCEPT one, REJECT two. To the right is an outline of Bowen's Reaction Series. A rock composed of the circled minerals formed by rapid cooling would be:

_____ 1. Andesite porphyry

_____ 2. Scoria

_____ 3. Obsidian

O
P
A
B

Ca
Ca/Na
Na

O
M
Q

CRITICAL REASONING PROBLEMS FOR WHICH YOUR INSTRUCTOR WILL HAVE TO SUPPLY ROCK SPECIMENS

Studying the names of rocks is no substitute for studying rocks themselves. This is because many rocks with the same name look superficially different, and you need practice looking at rocks, identifying texture, composition, and color and deciding what they are.

There are three strategies presented. Each uses the block diagram on page 54 showing a portion of the earth's lithosphere.

Strategy One

Ask your instructor to give you an unknown rock which might form at one of the positions on the block diagram. On a separate piece of paper
1. Identify the rock.
2. Write the letter of the position it would most likely be found.
3. Explain in writing why that rock must be found at that locality and no other.

Strategy Two

Ask your instructor to identify one location of the cross section and give you several unidentified igneous rocks, one of which could come from the identified location. On a separate piece of paper
1. Choose the one rock which would most likely come from the location.
2. Explain in writing why that rock must be found at that locality.
3. Explain in writing why each of the remaining rocks cannot come from the locality.

Strategy Three

Ask your instructor to give you several igneous rocks, each of which could come from a location on the cross section. On a separate piece of paper
1. Identify each igneous rock.
2. Identify the location of each rock on the cross section.
3. Explain in writing the process(es) of formation of each rock.

DECISION TREES FOR DISTINGUISHING AMONG IGNEOUS ROCKS
LISTED ON PAGE 30

Decision Tree for Pumice, Obsidian, Scoria, Vesicular Basalt

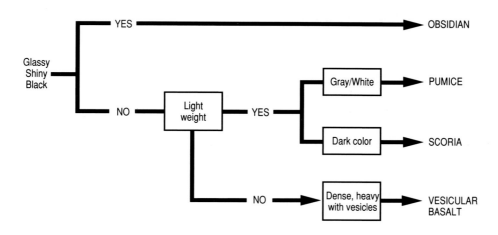

Decision Tree for Peridotite, Gabbro, Diorite

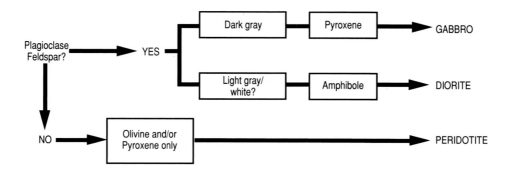

Decision Tree for Basalt, Obsidian, and Peridotite

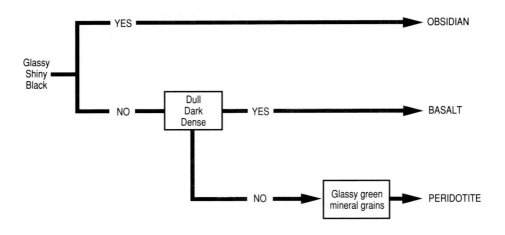

Decision Tree for Granite, Rhyolite Porphyry, and Granite

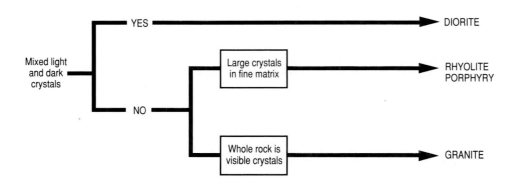

Preliminary to Sedimentary Rocks

PURPOSE

To understand sedimentary rocks, you must integrate the skills and knowledge gained in the *Minerals* and *Igneous Rocks* labs with knowledge of the processes which operate at the earth's surface. Upon completion of this laboratory, you will understand that sedimentary processes are systematic and that they result in predictable changes in the preexisting igneous rocks and in predictable changes in how a sediment evolves.

This Preliminary introduces the models and concepts which make sedimentary rocks much easier to learn and understand. By the time you enter the next laboratory, you should:

1. Know the processes by which sedimentary particles form.

2. Know the **simple, ideal model** for the origin of the most abundant kinds of sediment (rock)—quartz sand (sandstone), clay (shale), and calcite in solution (limestone).

3. Know how the processes of weathering, sorting, size reduction, and depositional environments alter sediments as they are transported downstream.

4. Know what is meant by **sediment maturity** and how it measures the evolution of sedimentary rocks.

INTRODUCTION

If you were given a randomly selected pile of sedimentary rocks and asked to place them correctly into groups of related rocks, without any help, you might become frustrated. Sedimentary rocks come in great variety. But they need not be frustrating. The key to understanding sedimentary rocks is to understand (1) the processes by which sedimentary particles form and (2) what happens to these particles as they are transported and deposited.

We begin with a simple, ideal model that assumes that these two sets of processes have been driven to completion—that everything that can be done to a sediment has been done. Under these conditions we find that regardless of how complex the starting materials are, the final end products are always the same. Knowing what these end products are, and why they are what they are, is the key to understanding and interpreting sedimentary rocks, since nearly all sediments on the earth are at some stage on the way to these end products. The essence of sedimentary rock identification and interpretation is to determine how close any particular rock is to the end member stage.

THE ORIGIN OF SEDIMENTARY PARTICLES

The rock-forming minerals found in igneous rocks appear to be hard and durable. It is difficult to imagine them breaking down (weathering) and being washed away. Yet if these minerals are left exposed at the earth's surface, that is just what happens. They are attacked by the gases and water of the atmosphere, plus naturally occurring acids and other chemicals, and decompose into new minerals which are very different from the igneous rock–forming minerals.

Most minerals are stable only under the conditions at which they form. This seems obvious: if we raise the temperature above the melting point, it melts. But the processes of decomposition occur just as surely if the mineral is exposed at the conditions of the earth's surface. The process just takes longer there.

Many kinds of minerals and rocks are continuously being exposed at the earth's surface. These may be igneous rocks brought to the surface by volcanic activity or igneous, sedimentary, and metamorphic rocks brought to the surface by mountain building. In each case, as soon as these rocks are exposed at the surface, their minerals, formed under different conditions, begin to decompose and form new sedimentary minerals which are stable at the earth's surface. The weathering, erosion, transportation by water, wind, or ice (glaciers), and deposition are all part of the study of sedimentary rocks.

THE SIMPLE, IDEAL MODEL ILLUSTRATING THE ORIGIN OF THE THREE MOST ABUNDANT KINDS OF SEDIMENT—QUARTZ SAND, CLAY, CALCITE

Of the eight rock–forming minerals in Bowen's Reaction Series (olivine, pyroxene, amphibole, biotite, plagioclase, orthoclase, muscovite, and quartz), only quartz does not weather. Quartz is released from the rock by weathering but remains in the environment as sand–sized grains. The remaining seven minerals weather into two new end products: *clay*, a mineral made of very small crystals stable at the earth's surface, and *calcite* ($CaCO_3$) (plus other minor constituents) in solution. If weathering goes to completion, then of all the rock-forming minerals only these three remain:

1. quartz sand
2. clay (such as kaolinite)
3. calcite ($CaCO_3$) in solution

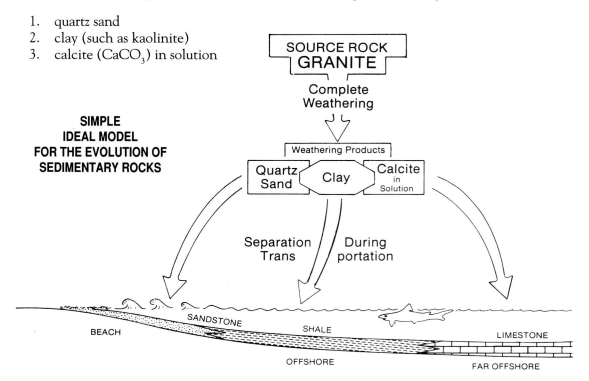

These three weathering products do not stay mixed in the **source land**[1] but are transported by streams and rivers down to the ocean, their final resting place. During transportation the quartz sand, clay, and calcite in solution are separated from each other because they do not all travel equally well.

The sand and clay, beginning as a poorly sorted mixture, are separated more and more as they travel downstream away from the source. Only at the beach are they finally separated completely. Waves crashing on the shore continuously stir up the sediment, but the quartz, being heavy, quickly settles back to the bottom. The clay, however, remains in suspension, drifts offshore into quiet water, and finally settles to the bottom. Now only the calcite in solution remains, and it finally is deposited in the far offshore, separated from both the quartz sand and clay. The calcite is deposited because plants and animals extract it from seawater and use it to build their skeletons. After death, their skeletons form the sediment.

These three end products compose most all sediments. When solidified, the three kinds of sediment form three kinds of sedimentary rock: quartz sandstone, shale, and limestone.

DEVIATIONS FROM THE SIMPLE, IDEAL MODEL

Between the unweathered source rock and the three end products of the simple, ideal model are many transition processes and sediments. Sedimentary rocks are not so confusing if you understand that most of them are in transition and that specific processes are driving them toward the ideal end members.

Weathering and the Source Rock   Weathering is not a simple process, and the initial weathering products are more diverse than are quartz sand, clay, and calcite in solution. Two kinds of weathering, and many kinds of weathering particles, can be found in the source land.

> *MECHANICAL WEATHERING*   Rocks are broken into smaller sizes, but the original mineral content does not change. These are called **lithic fragments** or **lithics** (lithic means "rock"). Initially, breakage produces angular lithic fragments, but these become rounded during transportation. Eventually, most lithic fragments are chemically weathered too.

> *CHEMICAL WEATHERING*   All minerals, except quartz, decompose to new minerals stable at the earth's surface. These include clay (many kinds), iron oxides (limonite, hematite), [2] and a great variety of minerals in solution. Minerals vary greatly in their resistance to weathering. Some are extremely stable and may survive nearly as long as quartz; others disappear very quickly. On Bowen's Reaction Series the minerals at the top weather most quickly, those at the bottom most slowly.

CONCLUSION ONE

Near the site of weathering, sediments begin as a mixture of any or all of the following: angular lithic fragments, feldspar, many other semistable mineral grains, clay, stable iron oxides, and minerals in solution—all mixed together.

---

[1] Any part of the earth that supplies sediment.

[2] The iron oxide minerals, although not abundant, are important because they make powerful stains (yellow, brown, green, purple, and red) and are very stable. As little as a few percent hematite can turn a rock vivid red. Gray and black are also common colors caused by decayed organic matter; it can mask colors caused by iron oxides.

Sorting   Large, heavy particles require more effort (greater water velocity) to transport than do small, light particles. In a stream,

1. Minerals in solution leave first, even if the water is only a trickle and nothing else moves.

2. Clay and iron oxides, the smallest sedimentary particles, move next; they are kept in suspension by gentle flow and travel almost as fast as the stream.

3. Sand grains move in water of moderate energy by bouncing and rolling along the bottom.

4. Larger mineral and lithic fragments finally begin to move only when water velocity is very high.

In a swift stream it may seem like most particles are moving quickly, but if water velocity slows even a little, the larger particles stop moving first and get left behind.

CONCLUSION TWO

Sedimentary particles are sorted downstream by water (and wind) according to size—large particles left behind upstream, and smaller particles moved farther downstream.

Rounding and Size Reduction   Angular sedimentary particles have not been transported very far from their source. Particles which are transported by running water have their sharp edges knocked off by abrasion with other particles and become ever more rounded. Particles also become smaller because of abrasion and breakage.

CONCLUSION THREE

Sedimentary particles increase in roundness and decrease in size downstream.

Depositional Environments   There are many depositional environments, and each has a unique set of processes which controls how sediment is transported and deposited. Swift-flowing rivers with great turbulence transport larger particles easily and carry away the finer particles, producing a coarse grained sedimentary rock. Deeper, gentler-flowing water transports only the finer particles producing a shale.

Depositional environments, like sediments, evolve in a systematic and predictable way downstream. The downstream sequence of environments is (see illustration, page 63):

Alluvial Fan ⇒ Braided River ⇒ Meandering River ⇒ Delta/Beach ⇒ Shelf ⇒ Submarine Fan ⇒ Basin.

In general, the total amount of energy working on a sediment decreases downstream, meaning large particles tend to be at the beginning in alluvial fans and braided rivers and smaller particles at the end of the sequence. Individual environments can be missing from the sequence, but the sequence is never rearranged so that an environment farther downstream comes before one farther upstream. Short systems, ones with many or most environments missing, are common in areas which are geologically very active. In addition to the environments just listed are other environments not listed. A few are shown on the drawing, but they are not essential to the sequence.

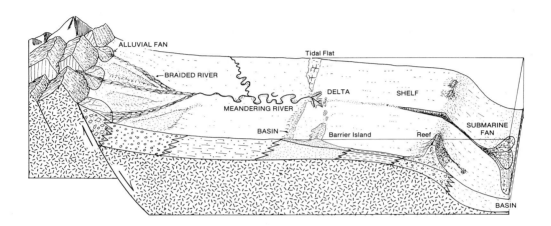

Just from studying sedimentary rocks, ancient depositional environments can be recognized. There are two principal kinds of evidence. First is sedimentary rock composition. Second are sedimentary structures (the patterns formed by sediments as they are deposited). We will occasionally need to refer to depositional environments in passing, but the study of sedimentary structures and environments is a specialized skill we cannot pursue here. What is important to recognize is that sedimentary rocks are not accidental, but are controlled by systematic laws and definite depositional processes.

CONCLUSION FOUR

The composition, size and shape, and sorting of a sediment are largely dependent on the depositional environment in which it was deposited. The characteristics of a sediment change in a systematic and predictable way downstream from environment to environment.

SEDIMENTARY ROCK CLASSIFICATION

No one feature can be used to classify sedimentary rocks. Almost every feature which can be used to classify one rock is also found in some other rock which formed in very different ways or under very different conditions. Sedimentary rocks form in so many different ways that it took geologists a long time to devise the classifications we now use.

Our classification groups sedimentary rocks into three categories based on their method of origin: **clastic, chemical,** and **biochemical.** This system puts together rocks which often look very different from each other and separates rocks which often appear quite similar. Nonetheless, what geologists want to understand are the *processes* by which rocks form, and this classification reflects that by grouping together sedimentary rocks which are process related.

The diagram below is derived directly from our simple, ideal model. Each end product in the simple, ideal model represents a class of sedimentary particles: quartz—all visible grains; clay—all clay-sized grains; calcite—all minerals in solution. Under each class in the diagram are listed other weathering products belonging to the class. The three classes of weathering products result in the three categories of sedimentary rocks.

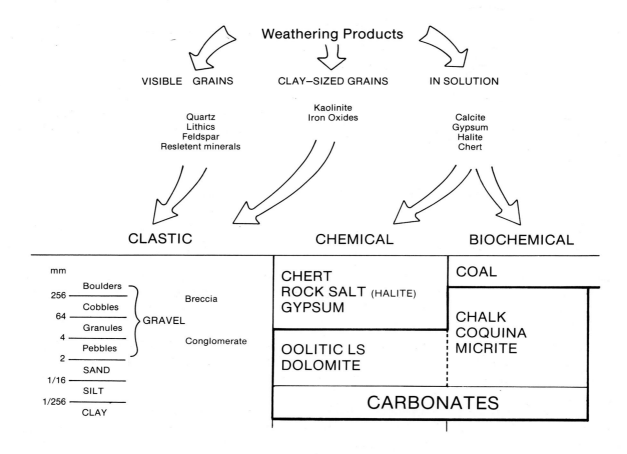

CLASTIC ROCKS   are made of the undissolved weathering products of preexisting rocks and are recognized, identified, and named by the size of the particles which form them. The weathering products are called **clasts.** The **Wentworth Scale** divides the clast sizes into the clay, silt, sand, and gravel categories in the lower left of the diagram. Despite great differences in superficial appearance, the rocks are relatively easy to identify once a few examples are studied.

In addition to this size name, it is common to describe the rock's composition too. For example, a rock with large amounts of feldspar is called an arkose breccia, or arkose conglomerate, or arkose sandstone, depending on the grain size and shape. The second-level classification in the lab will explore this.

CHEMICAL ROCKS form when the weathering products of preexisting rocks go into solution as dissolved minerals. Certain kinds of limestone (e.g., oolites) and dolomite ($CaMg(CO_3)_2$) are the most abundant chemical rocks. Although composed of the minerals calcite and dolomite, they don't look like the crystal forms you may have seen in the mineral lab. Both of these are readily identified by fizzing with weak hydrochloric acid (dolomite has to be powdered).

Halite and gypsum are usually identical with the mineral form, and frequently the mineral forms studied are sedimentary in origin. They precipitate from highly concentrated seawater in arid climates with high evaporation.

Chert usually forms from the recrystallization of some other form of silica, either skeletons of sponges or microorganisms, or volcanic ash.

BIOCHEMICAL ROCKS are formed from the skeletons of organisms. Most limestones are biochemical, although they intergrade with chemically formed limestone and dolomite. Typical examples are micrite ("lime mud", or clay-sized calcite crystals) and chalk. We will distinguish a number of biochemical limestones.

MEASURING THE MATURITY AND IMMATURITY OF CLASTIC SEDIMENTS

One of our goals with clastic rocks is to measure their progress toward the ideal, ultimate end product, quartz sand. Sediment maturity is the most important measure of this.

> **MATURITY** A relative measure of how extensively and thoroughly a clastic sediment (sand size and larger) has been weathered, transported, and reworked toward its ultimate end product, quartz sand.

Sediment maturity involves both composition and texture (grain size, roundness, and sorting).

A *compositionally* mature clastic sediment is one which has reached the ultimate end member of the ideal model, quartz sand. All other mineral grains, lithic fragments, and clay have been removed by weathering or sorting.

A *texturally* mature clastic sediment is one in which all the sediment grains have been worn by abrasion to perfect spheres and sorted perfectly to size (all particles at any one place are the same size).

For a sandstone to attain compositional and textural maturity is difficult and rare. It probably requires that a sediment go through several cycles of weathering, transportation, and deposition before compositional maturity is reached. Textural maturity requires conditions in the depositional environment which are uniform over long periods of time.

Descriptions of Different Sediment Maturity

Immature clastic sediments contain many of the following: angular fragments, many different minerals and lithic fragments, and many sizes of particles. Immaturity implies very incomplete weathering, short transportation, and rapid burial, like an alluvial fan.

Submature clastic sediments contain abundant quartz, but also lesser amounts of feldspar, lithics, silt, and clay. The sand-sized particles are beginning to round. Submaturity implies more thorough but still incomplete weathering, intermediate distances of transportation, and depositional environments where sorting is poor to moderate, like a river channel.

Mature clastic rocks contain quartz grains only, of sand size, with excellent rounding and sorting. Maturity of sandstones implies extensive and complete weathering, great distances of transportation (or numerous cycles of deposition, erosion, and transportation), and depositional environments of generally high energy, like a beach.

Thus, you can see all sedimentary rock processes are directed toward two goals: (1) the total weathering of all igneous (and metamorphic) rock minerals to their ultimate end members—quartz sand, clay-sized grains, and minerals in solution—and (2) the separation of these end members into their preferred environments of deposition. What makes these goals so fundamentally important is that, in the final analysis, they are the most stable states that can be attained at the earth's surface. But these goals are not easily achieved, and many intermediate sediments and rocks are present on earth. The key to understanding these intermediate rocks is to understand the processes of their evolution. With these concepts, the great diversity of sedimentary rocks you will study in lab is more easily understood.

Sedimentary Rocks

THIS LABORATORY ASSUMES YOU KNOW OR CAN DO THE FOLLOWING:

1. Distinguish between **mechanical** and **chemical weathering** and the products of each.

2. Explain the **simple, ideal model for the evolution of sedimentary rocks** and why it is important.

3. Know the processes which modify a sediment as it is transported downstream.

4. Know the difference between **clastic, chemical,** and **biochemical** sedimentary rocks.

5. Explain what **sediment maturity** is and what it is used for.

> If you do not know the answers to these find out now. Read the Preliminary to Sedimentary Rocks, your notes, your textbook, or ask your neighbor.

ORGANIZATION

There are two parts to this laboratory. A summary of the subjects in each part follows.

PART ONE—SEDIMENTARY ROCK MINERALS   This part must be done before sedimentary rocks can be identified. It uses the mineral identification keys in the *Minerals* lab, pages 11-12.

PART TWO—SEDIMENTARY ROCK CLASSIFICATIONS   There are three levels of classification presented here.

Level One begins with a key to identify the origin of a sedimentary rock (clastic, chemical, biochemical). Level One also has the simplest classifications for clastic (texture only, key on page 73), chemical (key on page 74), and biochemical rocks (key on page 75). It can be used alone, without moving on to the other levels of classification, and it has an exercise on hypothesis-test decision trees to learn difficult specimens.

Level Two has more precise classifications for clastic rocks (based on composition and texture, key on page 79) and carbonates (pages 86-87). Level Two clastics can follow Level One directly, or Level One clastics can be skipped as long as you know what clay, silt, sand, and gravel are. Level Two carbonates are composed of calcite ($CaCO_3$) and dolomite [$CaMg(CO_3)_2$](includes chemical and biochemical rocks). Level Two can be done without doing Level One. It has an exercise on the evolution and maturity of sediments, page 80 (including critical reasoning problems).

Level Three has two sections—(1) classifying clastic sedimentary rocks with ternary diagrams (page 89) and (2) questions and critical reasoning problems (page 93).

┌─ *Part One* ──────────────────────────────────┐

SEDIMENTARY
ROCK MINERALS

Most of the minerals you identified in *Igneous Rocks* are unstable at the earth's surface. A typical granite contains feldspar, quartz, and biotite. Of these, only quartz is stable indefinitely; the others begin to decompose into minerals which are stable at the earth's surface. Thus, before sedimentary rocks can be identified, some new minerals must be learned. There are two new categories of sedimentary minerals: minerals in solution and clay, and silt-sized minerals.

In addition to the new minerals are several minerals present in igneous (and metamorphic) rocks which are nearly as stable as quartz, although eventually they do weather, including magnetite, garnet, zircon, rutile, tourmaline, and others. Many of these minerals are rare enough that you will not have to study them here, although they will exist in small percentages among the fragments in the rocks you study.

New Minerals

Very Stable Minerals	Clay and Silt Sized	Minerals in Solution
1. Quartz	1. Kaolinite (a clay)	1. Calcite
2. Magnetite	2. Limonite	2. Dolomite
3. Garnet	3. Hematite	3. Halite
4. Zircon	4. Bauxite	4. Gypsum
5. Rutile		5. Chert
6. Tourmaline		

IDENTIFYING SEDIMENTARY MINERALS

☐ (1) Get a tray of sedimentary minerals from your instructor. Work together in groups of two or three.

☐ (2) Select a mineral from the tray and write its number (if available) on the chart on page 69. If the specimens are not numbered, just arrange them in a row.

☐ (3) Determine the physical properties of the mineral and list them on the chart.

☐ (4) Use the "Key to the Identification of Minerals" on pages 11-12 and identify the mineral.

☐ (5) After you have identified two or three minerals, ask your instructor to check them before identifying the rest to make sure you are working correctly.

☐ (6) When all the minerals are identified and you feel confident you know them,
    A. Scramble all the minerals. Can you still identify them?
    B. Exchange your tray for another and look at different specimens of the same minerals. Can you still identify them?

PHYSICAL PROPERTIES			SPECIMEN NUMBER	MINERAL NAME
HARDNESS	LUSTER	CLEAVAGE		
STREAK	COLOR	OTHER	_____	
HARDNESS	LUSTER	CLEAVAGE		
STREAK	COLOR	OTHER	_____	
HARDNESS	LUSTER	CLEAVAGE		
STREAK	COLOR	OTHER	_____	
HARDNESS	LUSTER	CLEAVAGE		
STREAK	COLOR	OTHER	_____	
HARDNESS	LUSTER	CLEAVAGE		
STREAK	COLOR	OTHER	_____	
HARDNESS	LUSTER	CLEAVAGE		
STREAK	COLOR	OTHER	_____	
HARDNESS	LUSTER	CLEAVAGE		
STREAK	COLOR	OTHER	_____	
HARDNESS	LUSTER	CLEAVAGE		
STREAK	COLOR	OTHER	_____	
HARDNESS	LUSTER	CLEAVAGE		
STREAK	COLOR	OTHER	_____	
HARDNESS	LUSTER	CLEAVAGE		
STREAK	COLOR	OTHER	_____	

Part Two

SEDIMENTARY ROCK CLASSIFICATION

LEVEL ONE: BASIC CLASSIFICATION

At this level we do two things. First, we divide rocks into groups based on method of origin: clastic, chemical, and biochemical. Second, we identify the major kinds of sedimentary rocks.

☐ (1) Get a tray of sedimentary rocks from your instructor. Work together in pairs.

☐ (2) DETERMINE THE ROCK'S ORIGIN.
  A. If your rocks are mixed (or if the rock is just one you picked up somewhere), begin with the key on page 72 and identify the rocks origin. Place the rocks in three piles by origin. Ask your instructor to check your determinations.
  B. If the rocks are already divided into clastic, chemical, and biochemical, go directly to instruction 3.

> **OBSERVE:** Many sedimentary rocks are not purely clastic, chemical, or biochemical. For impure specimens, which group a rock belongs to is a judgment call. It is important that you observe the rock carefully and make the best judgment you can before asking for help. Right or wrong, you will learn more from making your own decision before someone tells you how to do it.

☐ (3) IDENTIFY INDIVIDUAL ROCKS. Follow steps A–D for each specimen.

STEP A Use the data sheets on pages 82–85 to record the rock's properties.

STEP B Analyze each rock for the properties listed below. In the beginning some of the properties may be unfamiliar and difficult to recognize, but by the end of the laboratory, they will be almost second nature. Not every rock has every property.

Origin—Check the box: clastic, chemical, biochemical.
Grain Size—If it is composed of individual grains, record their size; a ruler is drawn on each data page and on the clastic identification key, page 73.
Grain Shape—If appropriate: angular? subangular? subrounded? rounded?
Hardness—If the rock is composed of individual particles, check their hardness.
  If the rock is uniform in composition check overall hardness: harder than glass? softer than glass? harder than fingernail? softer than fingernail?
Acid Reaction—Test the rock with weak HCL acid. Does it react vigorously? weakly? only when powdered? not at all?
Composition—This space is for the chemical/biochemical rocks. You may not be able to identify composition until you go through the keys.

> **OBSERVE:** Many clastic rocks have some carbonate contamination. Just because a rock reacts does not automatically make it chemical or biochemical. If a rock is suspicious, ask for help.

Maturity—Skip this space at Level One.

Color—Color is helpful in identifying some chemical and biochemical rocks, but use color with great caution. Just a few percent of some iron oxide can markedly change the color of a rock.

Other—Some rocks have very distinctive features, such as the salty taste of rock salt. Record in the space.

STEP C Using the keys on pages 73–75, identify the rock.

STEP D After you have identified two or three rocks, ask your instructor to check them to make sure you are on the right track. Then proceed with the rest of the identifications.

HYPOTHESIS–TEST DECISION TREES FOR IDENTIFICATION

(4) If available, pick out and lay in front of you the following set of sedimentary rocks. Construct decision trees (see instructions on page 13–14 in *Minerals* lab) to identify the following sets of sedimentary rocks.
  A. Chalk, gypsum, micrite
  B. Siltstone, sandstone, shale
  C. Sandy siltstone, silty sandstone, siltstone
  D. Chert, micrite, dolomite, gypsum

Our decision trees for these sets of rocks are at the end of the chapter. After you have drawn your decision trees check with ours, but please don't peek first. It will not help you to look at ours before you draw yours.

Note that there is no one way to draw decision trees. Your decision trees may be different from ours but if they work that is all that matters.

KEY FOR DETERMINING THE ORIGIN OF SEDIMENTARY ROCKS

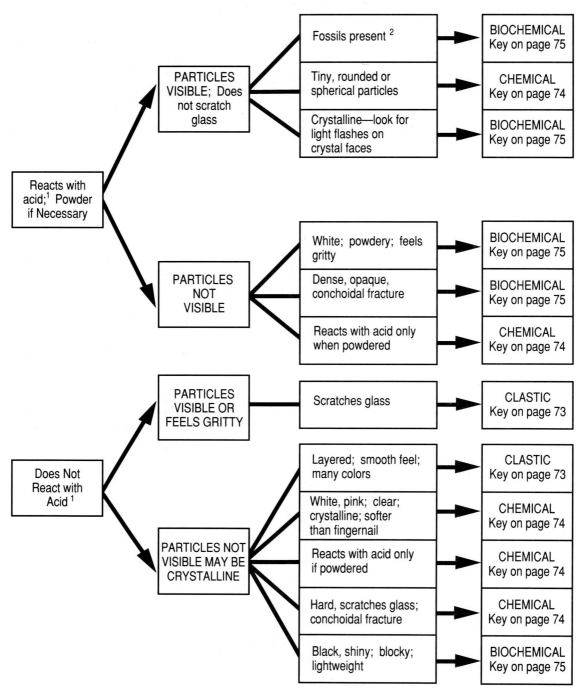

[1] Most rocks which react with acid are chemical or biochemical. However, some clastic rocks are contaminated with $CaCO_3$ and react with acid. These you must observe very critically. If the rock scratches glass, or has a clay residue after an acid reaction, it has some clastic particles in it. Ask your instructor for help on confusing specimens.

[2] Fossils may be present in clastic rocks too. If the rock scratches glass or leaves a powder residue after acid reaction, it is likely a clastic rock.

CLASTIC ROCKS BY TEXTURE
Level One Classification

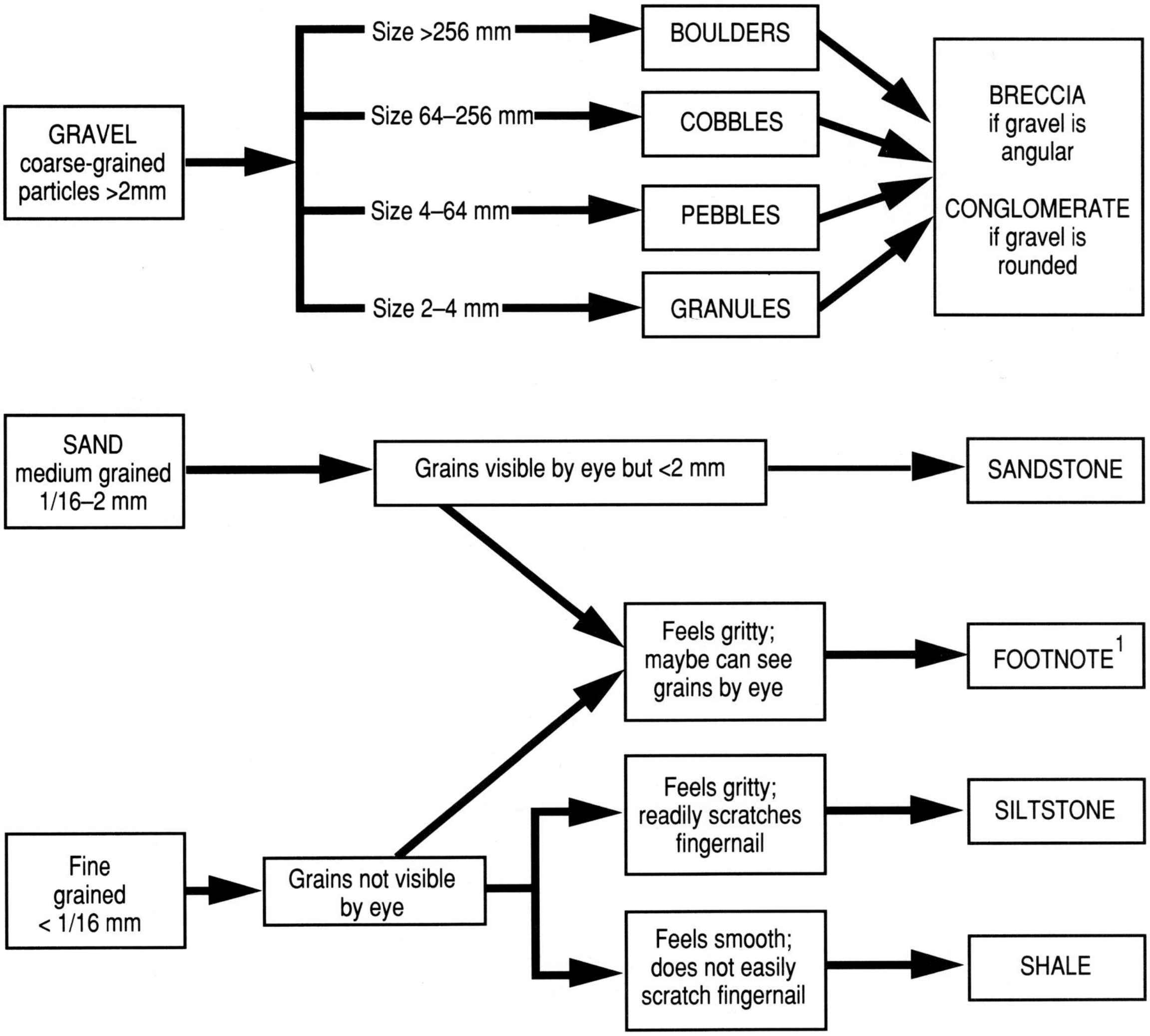

GRAVEL
coarse-grained
particles >2mm

Size >256 mm → BOULDERS

Size 64–256 mm → COBBLES

Size 4–64 mm → PEBBLES

Size 2–4 mm → GRANULES

BRECCIA
if gravel is
angular

CONGLOMERATE
if gravel is
rounded

SAND
medium grained
1/16–2 mm

Grains visible by eye but <2 mm → SANDSTONE

Feels gritty;
maybe can see
grains by eye → FOOTNOTE[1]

Fine
grained
< 1/16 mm

Grains not visible
by eye

Feels gritty;
readily scratches
fingernail → SILTSTONE

Feels smooth;
does not easily
scratch fingernail → SHALE

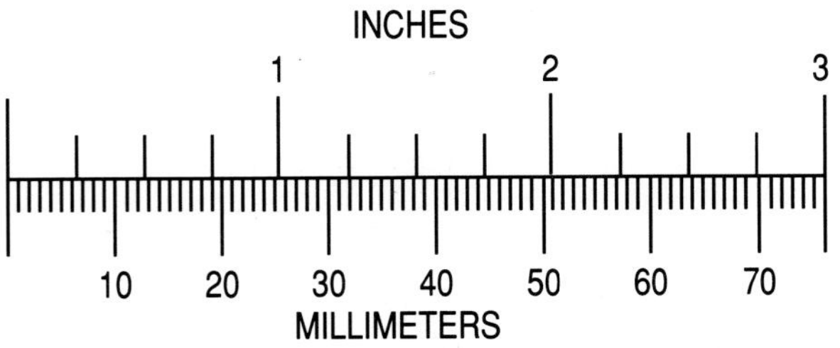

INCHES

1   2   3

10 20 30 40 50 60 70

MILLIMETERS

[1] These rocks can be difficult to classify. They may be fine-grained sandstones, or sandy siltstones, or sandy shales. Use a microscope. If the rock appears to be mostly tiny sand grains, call it a **fine-grained sandstone**. If the rock has lots of sand (>50%) embedded in silt or clay, call it a **dirty sandstone**. If the rock is mostly silt or clay with some sand (<50%), call it a **sandy siltstone** or **sandy shale**.

CHEMICAL SEDIMENTARY ROCKS
Level One Classification

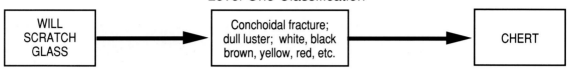

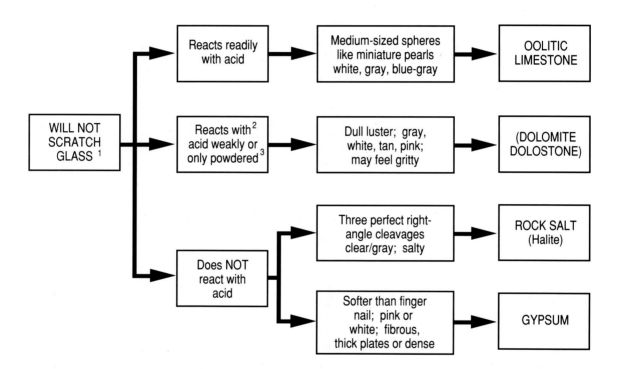

[1] Be cautious about contaminated specimens. Some rocks softer than glass may have some grains which will scratch glass. Run the hardness test carefully several times and observe carefully.

[2] Rocks may vary from pure limestone to pure dolomite with every intermediate mixture. The acid reaction varies from vigorous to weak across the spectrum. Observe how pure calcite reacts as a basis of comparison with rocks mixed with dolomite.

[3] You can powder dolomite by rubbing it on a streak plate until a small pile accumulates, or scrape the rock with a knife. Pure dolomite reacts very slowly so observe closely.

BIOCHEMICAL SEDIMENTARY ROCKS
Level One Classification

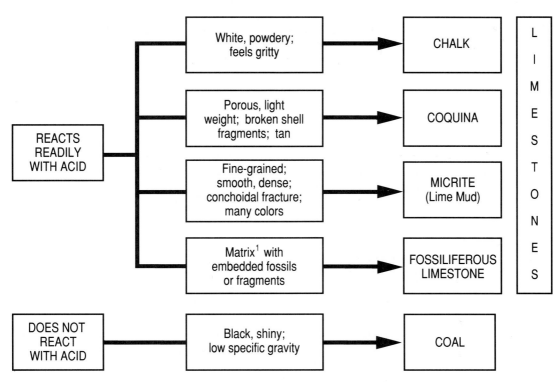

Note: All the rocks that react readily with acid are collectively known as *limestones*.

LEVEL TWO: CLASTIC CLASSIFICATION

In Level Two we do two things. First, we identify clastic sedimentary rocks by composition and texture. Second, we use our knowledge of composition to study how clastic rocks evolve.

In Level One classification, clastic rocks are identified by texture alone. A complete classification must include both texture and composition. But what should be included in composition? There is an almost unlimited variety of particles which could end up in a sedimentary rock! Consider, for example, any piece of an igneous, sedimentary, or metamorphic rock; any of the several thousand minerals which are now known; or any of the weathering products such as silt, clay, and iron oxides. A classification based on composition *could* become very complicated, and confusing, if all these were considered. But in most cases, composition is lumped into only four categories, as follows.

CATEGORIES OF CLASTIC SEDIMENTARY ROCK COMPOSITION

QUARTZ Since quartz, for all practical purposes, does not weather into anything else and will remain after everything else is weathered or sorted out, it is one of the most important of the four components of sedimentary rock composition.

---

[1]Matrix may be lime mud (micrite) or calcite crystals with cleavage faces.

MATRIX: **Silt and Clay**   Since all minerals other than quartz will eventually weather into silt or clay-sized particles, silt or clay is the second of the four components of sedimentary rock classification. Matrix is the finer material in which larger particles are embedded.

FELDSPAR   Feldspars are some of the most abundant minerals in the earth's crust. With only a few exceptions all igneous rocks have large amounts of feldspar, for example, calcium plagioclase in gabbro, and sodium plagioclase and orthoclase in granite. Sediments near the source land frequently have large percentages of feldspar, and they are important in sedimentary rock composition.

LITHIC FRAGMENTS   Very simply, if a clastic particle is not one of the first three, it is classified a **lithic fragment. Lithic** means "rock," and all mechanically weathered pieces of another rock, or nonfeldspar minerals weathered from a rock, are included here. Frequently they are small, dark in color, and difficult or impossible to identify specifically in hand specimen.

A sedimentary rock composed of only one of these four components is easy to identify, but most of the sedimentary rocks we find are mixtures of two or more of the four components. Still, even here the variety of sedimentary rock names is limited. The classification is straightforward and not at all difficult to remember.

LEVEL TWO IDENTIFICATION— CLASTICS AND THE EVOLUTION OF SEDIMENTARY ROCKS

☐   (1)   Get a tray of clastic sedimentary rocks from your instructor and the chart in the pocket at the back of the manual entitled, *"The Evolution of Sedimentary Rocks."* If your tray is mixed, remove all the clastic rocks for study. Work together in groups of two or three.

☐   (2)   ANALYZE INDIVIDUAL ROCKS.

STEP A   If you have already identified the clastic rocks at Level One (key on page 73), then you already know a lot about them and you just need to estimate compositional abundance. (If you have NOT done Level One go to step B.)

OBSERVE: Many sedimentary rocks are not purely clastic, chemical, or biochemical. For impure specimens, which group a rock belongs to is a judgment call. It is important that you observe the rock carefully and make the best judgment you can before asking for help. Right or wrong, you will learn more from making your own decision before someone tells you how to do it.

---

1.   Using a hand lens or microscope, identify which of the four kinds of particles are present in the rock: **quartz, silt/clay, feldspar, lithic fragments** (see two observe boxes on next page for help and hints).

2.   Record the composition on the charts on pages 82–85 *in order of abundance* putting the most abundant at the top.

3.   Go to instruction 3.

STEP B   If you have skipped the Level One identification for clastic rocks, analyze each rock for the properties listed on the data sheets, pages 82-85. In the beginning some of the properties may be unfamiliar and difficult to recognize, but by the laboratory's end they will be almost second nature. Not every rock has every property.

Origin—Clastic.

Grain Size—Record the grain size; a ruler is drawn on each data page and on the identification key, page 73.

Grain Shape—Angular? subangular? subrounded? rounded?

Hardness—If the rock is uniform in composition, check overall hardness: harder than glass? softer than glass? harder than fingernail? softer than fingernail?

Acid Reaction—Test the rock with weak HCL acid. Does it react vigorously? weakly? only when powdered? not at all?

Composition—Using a hand lens or microscope, identify which of the four kinds of particles are present in the rock: **quartz, silt/clay, feldspar, lithic fragments.**

OBSERVE: It will be much easier to distinguish rock fragments, quartz, and feldspar if you wet the rock with water. Identifying the particles may take some practice. Make your best judgments and then ask for help.

Use the percentage composition chart on page 78 to make abundance estimates. Observe the rock under a microscope and compare the abundance of quartz, feldspar, lithics present; estimate matrix last. Make two estimates, the first when the rock is dry and the second when it is wet with water. If these estimates do not agree, try to find a compromise which resolves the differences.

Record the composition on the charts on pages 82-85 *in order of abundance* from top to bottom.

OBSERVE: Estimating percentage abundance is a learned skill. Using a simple microscope there are no "right" answers, just estimates, some of which are better than others. Working with a partner, each of you make an estimate independently. Then compare; if you disagree by some large percent, explain to each other how you made your estimate and see if you can come to an estimate acceptable to both of you.

Maturity—Using the chart on page 78 determine the rock's maturity.

Color—You must use color with great caution. Just a few percent of some iron oxide can markedly change the color of a rock.

Other—If there is some distinctive feature record it in the space.

☐   (3)   Identify the rocks. Using the key on page 79, "Clastic Rock Identification: Texture and Composition," identify each rock and write its name in the space on the charts on pages 82-85. If two names seem to fit the rock, choose the one which is *most* likely.

☐   (4)   The evolution of sedimentary rocks. As each rock is identified, arrange it in its proper place on the foldout chart for *The Evolution of Sedimentary Rocks*. If more than one specimen seems to fit a space, arrange them in order downstream from most mature to least mature. The chart on page 78 summarizes the characteristics of immature, submature, and mature sedimentary rocks.

☐   (5)   Ask your instructor to check your rock identifications and the placement of the rocks on the evolutionary chart.

☐   (6)   Do the following section of critical reasoning problems on sediment maturity and the evolution of clastic rocks.

Applies to Clastic Sediments Only—Sand-Sized and Larger		
IMMATURE	SUBMATURE	MATURE
COMPOSITION Many minerals, especially feldspars and mica, and/or rock fragments	Quartz abundant, but other minerals or rock fragments common	Pure quartz (other minerals nonexistant or extremely rare)
— and —	— and/or —	— and —
TEXTURE Poorly sorted; many sizes	Sand common, but much silt and clay or quartz conglomerate	Sorting excellent (sand-sized only)
— and/or —		— and —
Angular fragments	Grains beginning to round	Rounding excellent
EXAMPLES Arkose/lithic breccia / Arkose/lithic conglomerate / Arkose/lithic sandstone	Quartz conglomerate / Wacke sandstones	Quartz sandstone

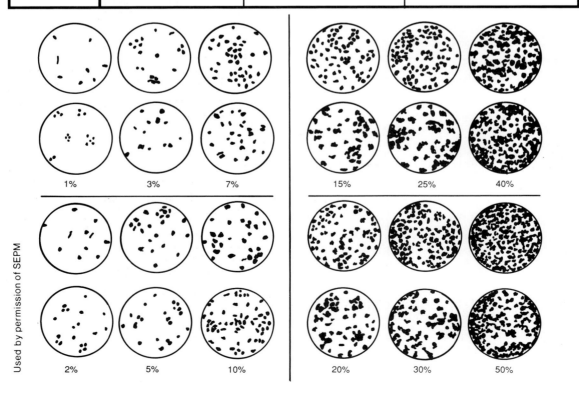

1%   3%   7%   15%   25%   40%

2%   5%   10%   20%   30%   50%

CLASTIC ROCK IDENTIFICATION: TEXTURE AND COMPOSITION

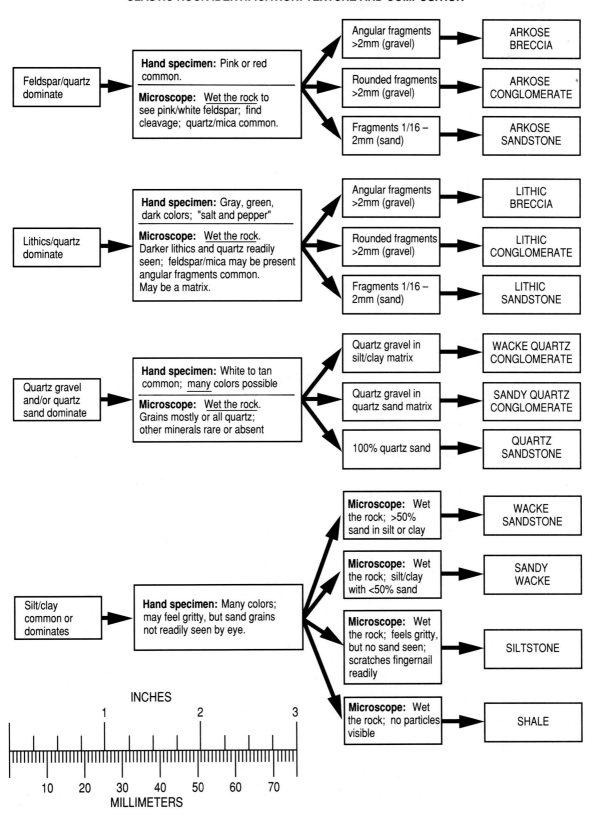

CRITICAL REASONING ABOUT SEDIMENT MATURITY AND THE EVOLUTION OF CLASTIC ROCKS

If all mountain building were to stop right now, in time every mountain would erode down to sea level. Then we would have the simple, ideal model for the evolution of sedimentary rocks. The only sediments left would be quartz sandstone, shale, and limestone.

But what happens before the simple, ideal stage is reached? How does a rock like granite become transformed into quartz sandstone, shale, and limestone?

The sediments evolve to the ideal state, beginning from a composition far from that end. **Maturity** is how the evolutionary progress of a sediment is measured.

Maturity measures how close a clastic sediment is to its ultimate end product, quartz sand. Immature and submature rocks contain minerals other than quartz, are poorly sorted (mixes of gravel, sand, silt, and/or clay), and have fragments which are more or less angular rather than rounded.

☐   (1)   Write answers to the following critical reasoning problems about the evolution of sedimentary rocks following these instructions:

         A.   For each problem there is a statement followed by several possible answers. For each possible answer you must indicate whether you REJECT it or ACCEPT it.

         B.   Logically and factually, in *writing,* explain your acceptance or rejection of *each* and *every* answer. There is NO CREDIT for a "right" answer, only for the analysis.

         C.   Only one answer will be *most* correct. Some answers will differ by subtle distinctions, but you must find the best answer and explain why the other answers are less correct.

         D.   You may discuss the problem with classmates, but when you write your analysis, it must be your own thinking, in your own writing.

PROBLEM NUMBERS ONE, TWO, THREE, AND FOUR

The first four problems on pages 93 and 96 should be done here if you are not going to do Level Three classification.

Problem Number Five

ACCEPT two, REJECT one. As a sediment is transported downstream the following changes occur:

    _____ 1.   Average grain size decreases, and angularity increases.

    _____ 2.   Grains become more rounded, and sorting increases.

    _____ 3.   Grain size becomes more uniform, and gravel decreases.

Problem Number Six

ACCEPT one, REJECT two. The drawing below shows a short system of depositional environments; that is, many of the depositional environments are missing and only an alluvial fan⇒braided river ⇒submarine fan is present. Compared to a complete system of depositional environments, you would predict the sediments in a short system would:

_____ 1. be coarser.

_____ 2. be more mature.

_____ 3. be better sorted.

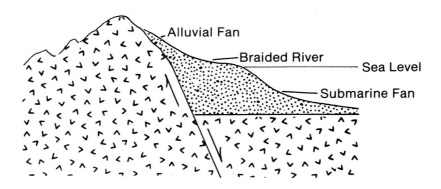

Problem Number Seven

ACCEPT one, REJECT two. For the same short system in Problem Number Six, compared to a complete system of depositional environments the sediments in a short system would:

_____ 1. be more quartz rich.

_____ 2. be more clay rich.

_____ 3. be more lithic or feldspar rich.

Problem Number Eight

ACCEPT one, REJECT two. If a source land were composed of clastic sedimentary rocks (rather than the igneous/metamorphic rocks as in the model for the evolution of sediments), then a sediment weathering from this source land would most likely be:

_____ 1. the same maturity as an igneous source land.

_____ 2. less mature than the igneous source land.

_____ 3. more mature than the igneous source land.

DATA SHEETS FOR LEVEL ONE CLASTIC, CHEMICAL, AND BIOCHEMICAL ROCKS
AND LEVEL TWO CLASTIC ROCKS

ROCK DESCRIPTION   SPECIMEN NUMBER   ROCK NAME

ORIGIN	GRAIN SIZE	GRAIN SHAPE
☐ Clastic ☐ Chemical ☐ Biochemical		HARDNESS
ACID REACTION	COMPOSITION ___ ___ % ___ ___ %	MATURITY
COLOR	___ ___ % ___ ___ %	OTHER

ORIGIN	GRAIN SIZE	GRAIN SHAPE
☐ Clastic ☐ Chemical ☐ Biochemical		HARDNESS
ACID REACTION	COMPOSITION ___ ___ % ___ ___ %	MATURITY
COLOR	___ ___ % ___ ___ %	OTHER

ORIGIN	GRAIN SIZE	GRAIN SHAPE
☐ Clastic ☐ Chemical ☐ Biochemical		HARDNESS
ACID REACTION	COMPOSITION ___ % ___ %	MATURITY
COLOR	___ ___ % ___ ___ %	OTHER

ORIGIN	GRAIN SIZE	GRAIN SHAPE
☐ Clastic ☐ Chemical ☐ Biochemical		HARDNESS
ACID REACTION	COMPOSITION ___ % ___ %	MATURITY
COLOR	___ ___ % ___ ___ %	OTHER

INCHES

1   2   3

10   20   30   40   50   60   70

MILLIMETERS

DATA SHEETS FOR LEVEL ONE CLASTIC, CHEMICAL, AND BIOCHEMICAL ROCKS
AND LEVEL TWO CLASTIC ROCKS

ROCK DESCRIPTION   SPECIMEN NUMBER   ROCK NAME

ORIGIN	GRAIN SIZE	GRAIN SHAPE
☐ Clastic ☐ Chemical ☐ Biochemical		HARDNESS
ACID REACTION	COMPOSITION _____ ___ % _____ ___ % _____ ___ % _____ ___ %	MATURITY
COLOR		OTHER

ORIGIN	GRAIN SIZE	GRAIN SHAPE
☐ Clastic ☐ Chemical ☐ Biochemical		HARDNESS
ACID REACTION	COMPOSITION _____ ___ % _____ ___ % _____ ___ % _____ ___ %	MATURITY
COLOR		OTHER

ORIGIN	GRAIN SIZE	GRAIN SHAPE
☐ Clastic ☐ Chemical ☐ Biochemical		HARDNESS
ACID REACTION	COMPOSITION _____ ___ % _____ ___ % _____ ___ % _____ ___ %	MATURITY
COLOR		OTHER

ORIGIN	GRAIN SIZE	GRAIN SHAPE
☐ Clastic ☐ Chemical ☐ Biochemical		HARDNESS
ACID REACTION	COMPOSITION _____ ___ % _____ ___ % _____ ___ % _____ ___ %	MATURITY
COLOR		OTHER

INCHES

1   2   3

10   20   30   40   50   60   70

MILLIMETERS

DATA SHEETS FOR LEVEL ONE CLASTIC, CHEMICAL, AND BIOCHEMICAL ROCKS
AND LEVEL TWO CLASTIC ROCKS

ROCK DESCRIPTION SPECIMEN NUMBER ROCK NAME

ORIGIN	GRAIN SIZE	GRAIN SHAPE
☐ Clastic ☐ Chemical ☐ Biochemical		HARDNESS
ACID REACTION	COMPOSITION ____ ___ % ____ ___ %	MATURITY
COLOR	____ ___ % ____ ___ %	OTHER

ORIGIN	GRAIN SIZE	GRAIN SHAPE
☐ Clastic ☐ Chemical ☐ Biochemical		HARDNESS
ACID REACTION	COMPOSITION ____ ___ %	MATURITY
COLOR	____ ___ % ____ ___ %	OTHER

ORIGIN	GRAIN SIZE	GRAIN SHAPE
☐ Clastic ☐ Chemical ☐ Biochemical		HARDNESS
ACID REACTION	COMPOSITION ____ ___ % ____ ___ %	MATURITY
COLOR	____ ___ % ____ ___ %	OTHER

ORIGIN	GRAIN SIZE	GRAIN SHAPE
☐ Clastic ☐ Chemical ☐ Biochemical		HARDNESS
ACID REACTION	COMPOSITION ____ ___ % ____ ___ %	MATURITY
COLOR	____ ___ % ____ ___ %	OTHER

INCHES

1   2   3

10   20   30   40   50   60   70

MILLIMETERS

DATA SHEETS FOR LEVEL ONE CLASTIC, CHEMICAL, AND BIOCHEMICAL ROCKS AND LEVEL TWO CLASTIC ROCKS

ROCK DESCRIPTION SPECIMEN NUMBER ROCK NAME

ORIGIN	GRAIN SIZE	GRAIN SHAPE
☐ Clastic ☐ Chemical ☐ Biochemical		HARDNESS
ACID REACTION	COMPOSITION ___ ___ % ___ ___ % ___ ___ % ___ ___ %	MATURITY
COLOR		OTHER

ORIGIN	GRAIN SIZE	GRAIN SHAPE
☐ Clastic ☐ Chemical ☐ Biochemical		HARDNESS
ACID REACTION	COMPOSITION ___ ___ % ___ ___ % ___ ___ % ___ ___ %	MATURITY
COLOR		OTHER

ORIGIN	GRAIN SIZE	GRAIN SHAPE
☐ Clastic ☐ Chemical ☐ Biochemical		HARDNESS
ACID REACTION	COMPOSITION ___ ___ % ___ ___ % ___ ___ % ___ ___ %	MATURITY
COLOR		OTHER

ORIGIN	GRAIN SIZE	GRAIN SHAPE
☐ Clastic ☐ Chemical ☐ Biochemical		HARDNESS
ACID REACTION	COMPOSITION ___ ___ % ___ ___ % ___ ___ % ___ ___ %	MATURITY
COLOR		OTHER

INCHES

1   2   3

10   20   30   40   50   60   70

MILLIMETERS

LEVEL TWO: CARBONATE ROCKS

In the simple, ideal model for the evolution of sedimentary rocks, we said that if weathering, transportation, and sorting go to completion, all that remains are three end members.

Clastic rocks

QUARTZ SANDSTONE ———→ SHALE ———→ LIMESTONE

Limestone and related rocks are explored here. Limestone is not a single rock but a group of related rocks, all composed of $CaCO_3$ and reacting with acid. In addition, limestone is closely related to dolomite $[CaMg(CO_3)_2]$. Because all these rocks have CO_3 in common, they are called the **carbonates.**

Lumping limestone and dolomite together ignores their chemical/biochemical differences, but it is not always easy to determine what part of a rock is biochemical and what part is chemical. The diagram in the Preliminary on page 64 shows which carbonates are generally chemical and which are biochemical.

The origin of dolomite is one of the major unsolved problems in geology. Debate has raged over whether dolomite is "primary" (precipitating directly from sea water) or "secondary" (limestones converted to dolomite after deposition). Simplistically, it appears that most dolomite is secondary in origin and forms in several ways, so any limestone can be converted to dolomite. In the classification here, it is the composition of the *original* limestone we are after, so if a rock has been converted to dolomite, look through that fact to the original rock.

Different Kinds of Carbonates   The name of every carbonate rock is a conjunction of two names, one describing the **Allochems,** the large pieces, the other describing the **Matrix.** There are four kinds of allochems:

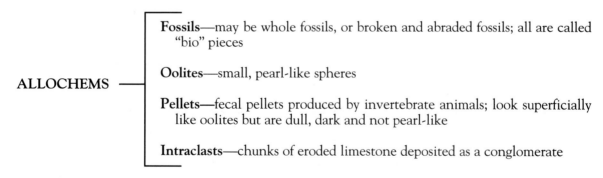

ALLOCHEMS

Fossils—may be whole fossils, or broken and abraded fossils; all are called "bio" pieces

Oolites—small, pearl-like spheres

Pellets—fecal pellets produced by invertebrate animals; look superficially like oolites but are dull, dark and not pearl-like

Intraclasts—chunks of eroded limestone deposited as a conglomerate

Micrite and sparite matrix

The matrix is of two kinds: MICRITE and SPARITE. Micrite is "lime mud," the dense, dull-looking lime sediment made of clay-sized crystals of $CaCO_3$. Micrite forms in the sea from the breakdown of skeletons. In the classification of carbonates we are concerned with whether micrite is present or absent as a matrix in the rock. If micrite is present during deposition, then it fills the spaces between the allochems, and the rock will be given a name which describes the allochems in a micrite matrix, for example, a rock with fossil fragments embedded in micrite is called a "biomicrite."

If, on the other hand, the depositional environment is such that no micrite is present, then the rock is composed only of allochems, held together by clear calcite crystals (called SPAR or SPARITE) precipitated by groundwater, after deposition, in the spaces between the allochems, binding them together.

A carbonate sediment composed only of allochems is like a sandstone with 100% sand and no silt or clay. All the micrite matrix has been washed away, as on a beach where no matrix can be deposited because the energy is too high. If we could see the sediment as it is being deposited, all the allochems would be loose, like pure sand or gravel. **SPAR** (or **sparite**) binds the allochems together. Spar is recognized by the calcite rhombohedral cleavage faces seen on broken surfaces.

Classification of Carbonates

The classification of carbonates using the allochem/matrix system is very simple and systematic. We combine the allochem name with the matrix name.

MATRIX

ALLOCHEMS		Micrite	Sparite
	Fossils	Biomicrite	Biosparite
	Oolites	Oomicrite	Oosparite
	Pellets	Pelmicrite	Pelsparite
	Intraclasts	Intramicrite	Intrasparite

This classification system has great flexibility and creativity, and you can build your own rock names easily. The name is built up by stringing together all the allochem names in order from least to most abundant and then adding the matrix name. For example, consider a rock like this:

Oolites	+	Fossils	+	Spar matrix	=	*Oo bio sparite*
(less abundant)		(most abundant)				

The name is written as one word, **oobiosparite.**

Another example:

Pellets	+	Oolites	+	Fossils	+	Micrite matrix	=	*pel oo bio micrite*
(least abundant)		(moderately abundant)		(most abundant)				

But what if there is both micrite and spar matrix? The system is the same.

Fossils	+	Spar matrix	+	Micrite matrix	=	*bio spar micrite*
(only allochem)		(least abundant)		(most abundant)		

LEVEL TWO IDENTIFICATION—CARBONATE SEDIMENTARY ROCKS

☐　　(1)　Get a tray of carbonate rocks from your instructor. If the tray is mixed, separate the carbonate rocks from the others.

☐　　(2)　From your carbonate rocks remove, if present, specimens of coquina, chalk, pure micrite, pure dolomite, and put them aside.

☐   (3)   Using a microscope or hand lens, identify the allochems, estimate their relative abundance, and list them on the charts on pages 88–89.

☐   (4)   Ask your instructor to check your identifications. If you have time get another tray with different specimens and practice some more.

Allochems in Order of Abundance

Least abundant   _____ , _____ , _____ , _____   Most abundant

Check the Box Which Describes the Matrix

☐ Micrite   ☐ Micsparite; mostly micrite, some spar   ☐ Sparmicrite; mostly spar; some micrite   ☐ Sparite

ROCK NAME: _____

Allochems in Order of Abundance

Least abundant   _____ , _____ , _____ , _____   Most abundant

Check the Box Which Describes the Matrix

☐ Micrite   ☐ Micsparite; mostly micrite, some spar   ☐ Sparmicrite; mostly spar; some micrite   ☐ Sparite

ROCK NAME: _____

Allochems in Order of Abundance

Least abundant   _____ , _____ , _____ , _____   Most abundant

Check the Box Which Describes the Matrix

☐ Micrite   ☐ Micsparite; mostly micrite, some spar   ☐ Sparmicrite; mostly spar; some micrite   ☐ Sparite

ROCK NAME: _____

Allochems in Order of Abundance

Least abundant   _____ , _____ , _____ , _____   Most abundant

Check the Box Which Describes the Matrix

☐ Micrite   ☐ Micsparite; mostly micrite, some spar   ☐ Sparmicrite; mostly spar; some micrite   ☐ Sparite

ROCK NAME: _____

Allochems in Order of Abundance

Least abundant _____ , _____ , _____ , _____ Most abundant

Check the Box Which Describes the Matrix

☐ Micrite ☐ Micsparite; mostly micrite, some spar ☐ Sparmicrite; mostly spar; some micrite ☐ Sparite

ROCK NAME: _____

Allochems in Order of Abundance

Least abundant _____ , _____ , _____ , _____ Most abundant

Check the Box Which Describes the Matrix

☐ Micrite ☐ Micsparite; mostly micrite, some spar ☐ Sparmicrite; mostly spar; some micrite ☐ Sparite

ROCK NAME: _____

Allochems in Order of Abundance

Least abundant _____ , _____ , _____ , _____ Most abundant

Check the Box Which Describes the Matrix

☐ Micrite ☐ Micsparite; mostly micrite, some spar ☐ Sparmicrite; mostly spar; some micrite ☐ Sparite

ROCK NAME: _____

Allochems in Order of Abundance

Least abundant _____ , _____ , _____ , _____ Most abundant

Check the Box Which Describes the Matrix

☐ Micrite ☐ Micsparite; mostly micrite, some spar ☐ Sparmicrite; mostly spar; some micrite ☐ Sparite

ROCK NAME: _____

LEVEL THREE: ADVANCED CLASSIFICATION OF CLASTICS

You have probably observed that the more closely you look at a rock the more there is to see, and the more difficult it is to identify. Yes, it may contain quartz and feldspar, but there is also some matrix (silt/clay) present, and perhaps a few rock fragments too. Perhaps you have exclaimed in frustration, "How do they expect us to identify this rock when it has a little bit of everything in it!" or "If the rock has both feldspar and lithic fragments, then it does not fit into any category we have and can't be classified!"

This is an extremely important insight. The closer we look at a rock (or any part of nature for that matter), the more there is to see. Furthermore, the more closely we look at a rock, the harder it is to decide exactly what the rock is—our knowledge increases but our confidence in accurately naming it decreases.

The Level Two classification is a simplification because it assumes two things: first, that the rock composition falls into simple categories, and, second, that the source lands weathering to produce the sediments are of only two kinds, either granitic continental or lithic volcanic. In the foldout chart on *The Evolution of Sedimentary Rocks* the first assumption is implied by the boxes containing rock names, and the second assumption by the presence of only two source lands.

Yet the chart on *The Evolution of Sedimentary Rocks* also implies that sedimentary rocks evolve in composition as they are transported, and not all sedimentary rocks will fall into our nice neat boxes—there will be transition rocks with a composition between one category and another. Source lands may also be complicated, for example, when a lithic volcanic mountain develops on a granitic continent. We will work with these situations when we discuss plate tectonics.

You will not be surprised, then, that geologists have developed a system for describing and naming all the intermediate rocks we find on the earth.

The Ternary Classification System for Sedimentary Rocks   The ternary diagram is the basis for most sedimentary rock classification. It has several distinct advantages. First, it easily incorporates three of the four components of sedimentary rock classification and in various technical forms can include more than that. Second, it allows description of all of the intermediate sedimentary rock compositions. Third, the naming of sedimentary rocks is not only more detailed, but they can be identified with as much or little precision as desired. Fourth, better interpretations can be made from a ternary diagram. And, last, the system once learned is very simple to use and modify for special purposes.

HOW A TERNARY DIAGRAM IS READ

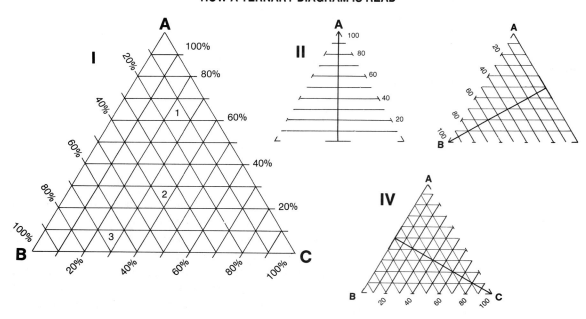

Reading a Ternary Diagram

See the ternary diagram (page 90) with the apexes labeled A, B, and C. A, B, and C can represent any composition. Let's take the diagram apart to see how it works.

The next panel (II) has only the skeleton of the triangle present as we concentrate on point A. Point A is at the end of the heavy vertical line. Along this line is indicated percent of A. A point plotted at the top of the line nearest A indicates 100% A. A horizontal bar at the bottom of the line (farthest from A) represents 0% of A. Any other percentage of A can be indicated by a bar appropriately located along the line between 0% and 100%.

The horizontal bars which indicate percent of A can be of any length since they remain the same distance from the bottom and top of the triangle. The bars are projected out to the right just as far as where the triangle side will be and percentage abundances written along the right side of the triangle (panel II). By doing this the right side of the triangle becomes the scale for percentage abundance of A. To be complete, the bars also extend to the left until they contact the left side of the triangle but no percent abundances are written there. In the final ternary diagram, the heavy vertical line is removed.

Point B is at the lower left apex of the triangle and is at the end of a heavy line as shown in panel III. This line can also be divided by bars to indicate percent abundance of B, and these bars extended out to the sides of the triangle. The left side of the triangle is chosen as the scale for percent abundance of B.

Point C is at the lower right apex of the triangle and is at the end of a heavy line as shown in panel IV. This line is divided by bars to indicate percent abundance of C, and these bars extended out to the sides of the triangle. Only the bottom of the triangle remains, and it is used for the scale of percent abundance C.

Overall, the ternary diagram is read **counterclockwise**, A on the right, B on the left, and C along the bottom.[1] The first panel (I) has some examples of points plotted with the following percentages:

1. 60% A,   20% B,   20% C = 100%
2. 30% A,   35% B,   35% C = 100%
3.   0% A,   70% B,   30% C = 100%

☐   (1)   On the same ternary diagram in panel I, try plotting the following points.

4. 80% A,   10% B,   10% C = 100%
5. 30% A,   70% B,   0% C = 100%
6. 40% A,   20% B,   40% C = 100%

☐   (2)   Check your plotted points with your instructor.

The Ternary Classification of Clastic Rocks

Consider the classification for clastic rocks on page 92. It uses two ternary diagrams: a *composition diagram* on the left, and a *texture diagram* on the right. The diagrams may be used separately or jointly, as explained below. Each ternary diagram is divided into fields and the names for the rocks in each field shown. The names are variations on names used at Level Two, key page 79.

The basic fields in the ternary diagrams are based on practical experience, but each field could be further subdivided as many times as wanted, and a name assigned to each field. A ternary diagram with many subfields allows very precise identification but becomes complicated. We will use the system in the simplest way in the examples below; ask your instructor for help if you want to create more precise names.

---

[1] The ternary diagram could be made to read **clockwise,** and in some areas of geology, such as soil studies, it is read that way.

Although the classification in the figure is conventional, it is not law. A geologist may devise a classification based on any ternary diagram she or he wants, making the system flexible and powerful. One advantage of the system is any classification is self-evident in the ternary diagram.

TERNARY CLASSIFICATION OF CLASTIC ROCKS

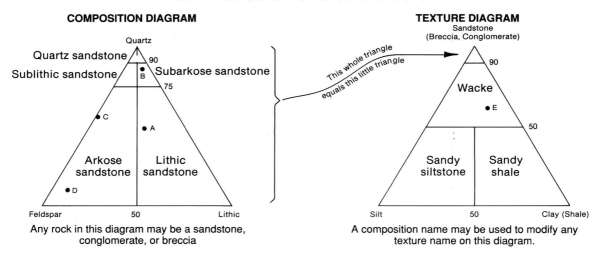

COMPOSITION DIAGRAM

Any rock in this diagram may be a sandstone, conglomerate, or breccia

TEXTURE DIAGRAM

A composition name may be used to modify any texture name on this diagram.

Reading the Composition Diagram   This left-hand composition diagram (above) is a **QFL (quartz/ feldspar/lithic)** diagram. It is standard to place quartz at the top, feldspar at the lower left, and lithics at the lower right. Rocks with over 90% quartz are quartz sandstones, or quartz breccias, or quartz conglomerates. (From now on when we say sandstone, understand it also means "or breccia or conglomerate.") Rocks with mostly quartz but 10 to 25% feldspar or lithics are subarkosic or sublithic sandstones. Rocks with over 25% feldspar or lithics, and the remainder quartz, are arkosic or lithic sandstones.

An example: Point A plotted in the composition diagram above is a rock with a composition of 50% quartz, 20% feldspar, and 30% lithic fragments. It is a **lithic quartz sandstone** because lithics and quartz are most abundant. If quartz is most abundant it is often omitted from the name—**lithic sandstone.**

In a similar way the following compositions in the ternary diagrams are named.

A: 50% Q,   20% F,   30% L = 100%   lithic (quartz) sandstone
B: 85% Q,   5% F,   15% L = 100%   sublithic (quartz) sandstone
C: 60% Q,   40% F,   0% L = 100%   arkosic (quartz) sandstone.
D: 10% Q,   80% F,   10% L = 100%   arkosic sandstone

Reading the Texture Diagram: This is for sandstones, or conglomerates, or breccias with lots of matrix (silt/clay).

The very top triangle of this ternary diagram is equivalent to the entire composition diagram on the left, that is, rocks with over 90% sand (gravel). The remaining fields include rocks with more and more matrix. **Wackes** are an important and common category; they are sandstones with 10 to 50% matrix (i.e., are texturally immature, poorly sorted). The sandy silts and sandy shales are over 50% matrix but may have from 0 to 49% sand (or gravel).

The composition and texture diagrams may be used at the same time to give a complete rock name. For example, point E in the texture diagram above is a wacke, just meaning it has sand with a clay matrix. But the sand portion alone could also be plotted on the composition diagram. If the sand portion of E were plotted on the composition diagram at the same spot as B, then E would be:

E: 40% matrix and 60% sand (which is 85% Q, 5% F, 15% L) = 100%
      The rock is a lithic wacke.

CLASSIFYING CLASTIC SEDIMENTARY ROCKS WITH A TERNARY DIAGRAM

☐   (1)   Get a tray of clastic sedimentary rocks from your instructor.

☐   (2)   On page 78 there is a percentage composition chart. Observing the rock under a microscope, compare the abundance of quartz, feldspar, and lithics present; estimate matrix last. Make two estimates, the first when the rock is dry and the second when it is wet with water. If these estimates do not agree, try to find a compromise which resolves the differences.

☐   (3)   Fill out the charts on pages 94–95 and using the ternary diagrams name the rock.

☐   (4)   Have your instructor check your identifications.

QUESTIONS AND CRITICAL REASONING PROBLEMS

Hidden within the sedimentary rock classifications you have studied are some very important concepts about science, and about what we ultimately can know. The following problems explore these concepts.

☐   (1)   For the critical reasoning problems follow the instructions on page 80.

☐   (2)   Write answers to all questions on a separate sheet of paper.

Problem Number One

ACCEPT two, REJECT one. What's in a name? Two people examine the same sedimentary rock. The first just calls it a "wacke" (a name meaning it has matrix and sand); the second names it a "subarkosic lithic quartz wacke" (meaning it has less than 25% feldspar, more than 25% lithics, 10 to 50% matrix, and the rest quartz).

_____ 1.   It is easier to identify a rock as a "wacke" than it is to identify it as a "subarkosic lithic quartz wacke."

_____ 2.   We are more confident the name "subarkosic lithic quartz wacke" is accurate (correct) than we are that "wacke sandstone" is accurate.

_____ 3.   The name "subarkosic lithic quartz wacke" carries more information (is more precise) than does the name "wacke."

TERNARY CLASSIFICATION OF CLASTIC ROCKS DATA SHEETS

Composition	Description Dry (use microscope)	Identify apexes and plot composition
_____ % Quartz _____ % Feldspar _____ % Lithic _____ % Matrix	Description Wet (use microscope)	
_____ Total Percent	Rock Name:	

Composition	Description Dry (use microscope)	Identify apexes and plot composition
_____ % Quartz _____ % Feldspar _____ % Lithic _____ % Matrix	Description Wet (use microscope)	
_____ Total Percent	Rock Name:	

Composition	Description Dry (use microscope)	Identify apexes and plot composition
_____ % Quartz _____ % Feldspar _____ % Lithic _____ % Matrix	Description Wet (use microscope)	
_____ Total Percent	Rock Name:	

Composition	Description Dry (use microscope)	Identify apexes and plot composition
_____ % Quartz _____ % Feldspar _____ % Lithic _____ % Matrix	Description Wet (use microscope)	
_____ Total Percent	Rock Name:	

TERNARY CLASSIFICATION OF CLASTIC ROCKS DATA SHEETS

Composition	Description Dry (use microscope)	Identify apexes and plot composition
_____ % Quartz _____ % Feldspar _____ % Lithic _____ % Matrix _____ Total Percent	Description Wet (use microscope) Rock Name:	

Composition	Description Dry (use microscope)	Identify apexes and plot composition
_____ % Quartz _____ % Feldspar _____ % Lithic _____ % Matrix _____ Total Percent	Description Wet (use microscope) Rock Name:	

Composition	Description Dry (use microscope)	Identify apexes and plot composition
_____ % Quartz _____ % Feldspar _____ % Lithic _____ % Matrix _____ Total Percent	Description Wet (use microscope) Rock Name:	

Composition	Description Dry (use microscope)	Identify apexes and plot composition
_____ % Quartz _____ % Feldspar _____ % Lithic _____ % Matrix _____ Total Percent	Description Wet (use microscope) Rock Name:	

Problem Number Two

ACCEPT one, REJECT two. If you understand the issues dealt with in Problem Number One, then we can say that, in principle,

_____ 1. An identification which carries a lot of information is more likely to be wrong than one which carries less information.

_____ 2. A less complex classification is more useful in a scientific study.

_____ 3. A precise identification is also a more accurate identification.[1]

Problem Number Three

Write a statement, both precise and accurate, which describes the relationships between the amount of information a scientific statement carries and its accuracy, precision, and vulnerability.

How confident are you that the relationships you describe are universal, applying to all forms of human knowledge?

Problem Number Four

Write a statement giving your opinion of the following question: In principle, do you believe it is better to have more information about something and be less confident it is correct, or to have great confidence, but less information? Logically justify your position and support it with an example.

Problem Numbers Five – Eight

These problems are on pages 80 and 81 and should be done here if you have not done them under Level Two classification.

Problem Number Nine

On page 97 is a ternary diagram. Note that it is different from any we have used so far; feldspar and lithics are plotted together in the lower left and matrix in the lower right; quartz is still at the top.

Get a tray of clastic sedimentary rocks from your instructor and using the ternary diagram do the following:

1. Plot the position of your samples on the ternary diagram.
2. Connect the plotted samples with arrows, beginning with the least mature and progressing to more mature. This would be from left to right on the "Evolution of Sedimentary Rocks" chart.
3. Describe in writing the trends in composition from sample to sample.
4. Write a statement explaining what maturity means on a ternary diagram.

---

[1] Precision and accuracy must be used very precisely and accurately. Imagine a target. If all the arrows hit the *same* spot, that is **precision**, even if they are far away from the center. **Accuracy** is if all the arrows cluster around the exact center of the target, no matter how scattered they are.

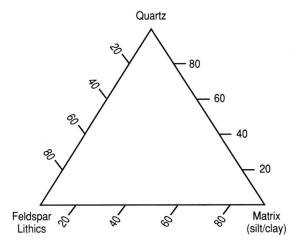

Problem Number Ten

Get the chart titled "Evolution of Sedimentary Rocks" from the pocket at the back of the manual. Arrange the rocks you have identified on the chart. There should be specimens which do not fit into the neat boxes on the chart. Arrange the rocks in the best order you can.

1. If you have specimens which in your opinion do not fit *this* chart very well, or at all, explain in writing why they do not fit.

2. Devise a chart which will fit the specimens you have.

DECISION TREES FOR DISTINGUISHING AMONG SEDIMENTARY ROCKS ON PAGE 71

Decision Trees for Chalk, Gypsum, and Micrite

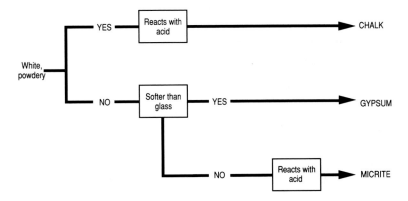

Decision Trees for Shale, Sandstone, and Siltstone

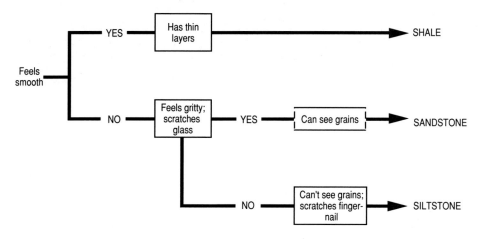

Decision Trees for Silty Sandstone, Sandy Siltstone, and Siltstone

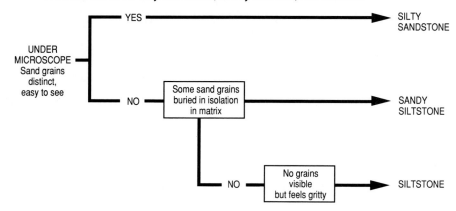

Decision Tree for Chert, Dolomite, Micrite, and Gypsum

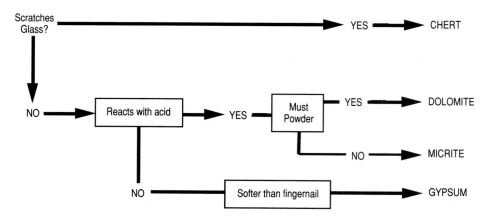

Preliminary to Metamorphic Rocks

PURPOSE

Metamorphic rocks are formed by the alteration of any preexisting rock by processes of heat, pressure, and/or hot fluids. To understand metamorphic rocks, you must apply all your knowledge of igneous and sedimentary rocks to imagine how they will change when subject to new metamorphic conditions. Despite the variety of metamorphic rocks, metamorphic processes are not random or haphazard but cause predictable changes in the preexisting (**parent**) rocks.

The premise here is that just as understanding how igneous and sedimentary rocks form and evolve makes those rocks easier to learn and understand, understanding metamorphic rock processes will make metamorphic rocks easier to learn and understand. This preliminary introduces some terms and concepts which will be necessary to do the laboratory which follows. By the time you enter the next laboratory, you should:

1. Know what is meant by the **geothermal gradient** and **metamorphic grade.**

2. Understand the concept of **metamorphic facies.**

3. Be able to list and briefly describe the four kinds of metamorphism, **regional, blueschist, eclogite,** and **contact/hydrothermal,** and locate them on a pressure/temperature diagram.

4. Know what is meant by the term **fabric** and the difference between **granular, foliated,** and **mineral banding** fabric.

INTRODUCTION

In the preliminary to sedimentary rocks, we introduced the principle that most minerals and rocks are stable only under the conditions at which they form. Raise or lower temperature and/or pressure, and they become unstable and transform into new minerals and rocks which are stable under the new conditions. So most metamorphism is the alteration of a rock through heat and pressure, but short of melting. The original rock prior to metamorphism is called the **parent rock.** A parent rock can be any preexisting rock, including igneous, sedimentary, and other metamorphic rocks.

Metamorphism is a natural response to burial since as a rock is buried deeper and deeper, both the pressure and the temperature affecting it increase. Temperature goes up because the core of the earth is hot enough to be molten and the closer a rock gets to it, the hotter it gets. Pressure goes up with burial because the deeper a rock is buried, the more overlying rock weighs down on it.

Metamorphic rocks are abundant in the earth, very often beautiful, and used as building stone. To the geologist, however, they are useful in the same way igneous rocks are. Because most metamorphic rocks form deep inside the earth where we can never go, when they are exposed at the earth's surface, then we can learn a lot about ancient conditions and geological processes. This discussion will give you a theoretical framework to understand how metamorphic rocks may be interpreted. First, the concepts of geothermal gradient, metamorphic grade, and metamorphic facies have to be introduced.

SOME BASIC METAMORPHIC CONCEPTS

The Geothermal Gradient If we could take an average spot on the earth, a place where no tectonic activity is occurring, and dig a mine deeper and deeper, the temperature would go up systematically as we get closer to the earth's molten core. This increase in temperature is known as the **geothermal gradient** and is plotted on the figure below. Temperature increases across the top left to right, as if one were getting closer to a furnace.

T-P DIAGRAM SHOWING GEOTHERMAL GRADIENT AND DIFFERENT TYPES OF METAMORPHISM

Temperature in Centigrade

But, of course, not only temperature goes up with depth but also pressure because the deeper you go the more rock is piled up overhead. Pressure increases down the side of the diagram, as if one were descending into the earth. The figure is, thus, a pressure-temperature diagram (or a P-T diagram). In geology, pressure is measured in **bars.** A bar is about 1 atmosphere of pressure, which is near 14.5 pounds per square inch at the earth's surface. A kilobar is 1,000 bars, and the pressure scale is measured in kilobars. Beside the kilobar scale is a depth scale corresponding to the increasing pressure. The depth scale is an ideal, just the pressure which would result from piling up rocks that high. In many places on the earth pressure can increase much faster than normal, while temperature remains low, such as at convergent plate boundaries (see below).

Metamorphic Grade—Measuring the Intensity of Metamorphism One of the geologists' most important questions is, "How intense was the metamorphism and how did it vary?" Conceptually it is straightforward. The intensity is determined by how close the country rock is to the source of heat and pressure. For example, deep in the core of a mountain, close to an igneous batholith, metamorphic processes are very intense. Sometimes the parent rock even melts. The farther one moves away from the batholith, the more the intensity of metamorphism declines, until finally the parent rock is unaltered. Another way of expressing this is, the intensity of alteration of the parent rock by the metamorphism has the potential to act like a barometer measuring pressure and a thermometer measuring temperature at various locations in the earth.

Metamorphism may result from several combinations of heat, pressure, and hot fluids, but in each case the intensity of metamorphism is one of the first things we want to learn. **Metamorphic grade** is a measure of how much a parent rock has been altered to become a metamorphic rock. The metamorphic grade for each kind of metamorphism is measured in its own way, as discussed shortly, but in general we speak of low-, medium-, and high-grade metamorphism.

Metamorphic Facies   This is perhaps the most central and important concept for understanding metamorphic rocks. The facies concept is not difficult, but you need to understand it clearly and unambiguously.

METAMORPHIC FACIES   A group of rocks that has been metamorphosed under similar temperature and pressure conditions

Note, it is the temperature and pressure conditions which are most critical in this definition. For a geologist it is the temperature and pressure conditions existing at the time of metamorphism that she or he wants to know.

Metamorphic facies, therefore, *do not* depend on the composition of the parent rock. And metamorphic facies are not any specific kind of rock. For example, a shale, a basalt, and a granite can all undergo greenschist facies metamorphism, and the resulting rocks will look nothing alike, but because they have all been subject to the same temperature and pressure conditions they all belong to the same metamorphic facies.

Several kinds of metamorphic facies can be defined depending on the combination of temperature, pressure, and hot fluids the parent rock was subjected to. The various kinds of facies are listed below. Also note the paths each kind of metamorphism takes across the temperature-pressure diagram.

1. **Regional metamorphism** (steadily increasing temperature and pressure), greenschist, amphibolite, granulite facies
2. **Blueschist facies metamorphism** (low-temperature, high-pressure)
3. **Eclogite facies metamorphism** (high-temperature, very-high-pressure)
4. **Contact/hydrothermal facies metamorphism** (high-temperature, low-pressure, usually with hot, chemical-rich fluids)[1]

Each of these kinds of metamorphism can produce a variety of metamorphic rocks, and in detail it can be very technical. In the lab, however, we will concentrate on the most basic and common kinds of metamorphic rocks and the facies in which they occur. For most kinds of metamorphic facies, you will just learn one or two typical rock examples. We believe it is centrally important, however, that you understand the larger framework into which each metamorphic rock fits, and this includes, in addition to the facies concept, the relationship between tectonics and metamorphism.

[1] This includes zeolite, hornfels, and prehnite-pumpellyite facies metamorphism. For our purposes we will lump these facies together. Ask your instructor if you want a more detailed explanation.

TECTONICS AND METAMORPHISM

What makes the metamorphic facies concept so powerful and important is that each kind of facies is directly related to a particular tectonic regime. Imagine, for example, a regime where pressure is low, but the temperature is high. Or a place where the pressure is high, but the temperature is low. Or a place where both temperature and pressure increase together from low to high levels. Or a place where both temperature and pressure are very high. As with igneous and sedimentary processes, each kind of metamorphism always occurs in specific places for specific reasons. A convergent plate boundary offers an ideal model in which virtually all of these combinations of temperature and pressure occur and can be explained in terms of large scale earth processes.

Model of a Convergent Plate Boundary   Below is a cross section of a portion of the earth's crust showing a convergent plate boundary.

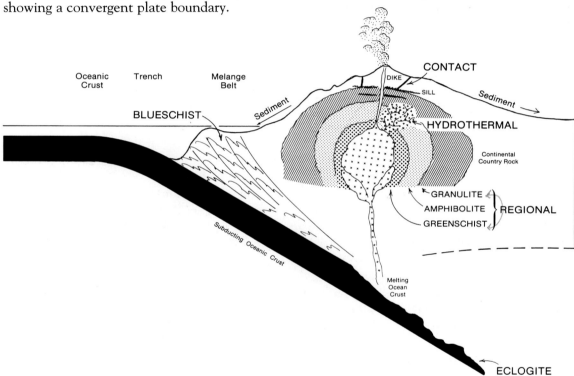

At a convergent boundary two of the earth's lithospheric plates are moving toward each other on a collision path. Inevitably, one or the other of the two plates will have to yield or get out of the way of the other. In the cross section the plate on the right is made of low-density granite (a continent) while the plate on the left is made of high-density mafic igneous rock (oceanic crust). It is the high-density mafic oceanic crust which gives way (or decouples) and dives into the earth under the edge of the continent. As the two plates continue to move together, the oceanic crust continues to slide into the mantle; that is, the oceanic crust is subducting.

A subducting oceanic plate sets in motion a series of processes, each of which is related to one or more kinds of metamorphism. We will begin by describing the processes along the convergent plate boundary and then tying the different kinds of metamorphism into those processes.

First, the oceanic plate as it subducts drags part of the earth's surface down with it creating a trench in the bottom of the ocean about 2 kilometers deeper than normal ocean depth. This trench will collect sediments eroding from the continent.

Second, the normally cold oceanic crust begins to heat up as it subducts. At about 120 kilometers down its upper portion is hot enough to begin to melt, releasing intermediate to felsic magma. The magma, being hot, fluid, and of lower density than the surrounding rock, rises toward the earth's surface intruding into the overlying continental country rock. Eventually the magma ponds as large batholiths somewhere below the surface, perhaps 10 to 20 kilometers down. From the batholiths some magma rises along volcanic pipes and dikes to the earth's surface and forms volcanoes.

Third, all the heat rising from the batholiths, and the compression of the two plates coming together, causes the crust to buckle upward and form a mountain range along the edge of the continent. These conditions create two kinds of pressure which affect the metamorphism. The first pressure is **hydrostatic,** that is, the pressure caused just by burial; it is directed equally in all directions. The second pressure is **compressional** or **directed,** that is, it squeezes the rock in one way.

Fourth, sediment eroding from the mountains flows away from the mountain in both directions, some onto the continent and some into the ocean basin. The sediment on the continent can just sit there, but the sediment which flows into the oceanic trench gets dragged down along the subducting plate because the plate is like a conveyor belt continuously moving into the earth. The sediment going down along the subducting plate gets badly sheared and deformed, and is called **melange** (French for mixture or jumble).

These processes occurring at a convergent plate boundary result in many kinds of metamorphism. Each kind of metamorphism is briefly discussed below with an explanation of where it fits into the model of an oceanic plate subducting under the edge of a continent.

Blueschist Facies Metamorphism   This metamorphism (low-temperature, high-pressure) is called blueschist because in outcrop the rock has a blue color caused by new high-pressure metamorphic minerals, such as glaucophane. Blueschist metamorphism occurs in the sediments (melange) being dragged down into the earth along the subducting oceanic crust. Blueschist metamorphism is low temperature because the melange is deposited deep in the cold ocean, on a crust of cold, mafic oceanic crust. Blueschist metamorphism is high pressure because the melange is quickly carried deep into the earth into a region of high pressure along the subducting slab.

Eclogite Facies Metamorphism   Eclogite (high-temperature, very-high-pressure) is a rock composed dominantly of pyroxene and red garnet (visually a very striking rock) and results from the high-temperature and high-pressure metamorphism of mafic igneous rocks. In the model it occurs as the unmelted portion of the subducting oceanic crust reaches great depths in the upper mantle. At these depths most of the rocks of mafic composition forming the upper mantle exist as eclogite.

Regional Metamorphism   By far the most common type of metamorphism is regional metamorphism (steadily increasing temperature and pressure). Regional metamorphism occurs in the rocks surrounding the intermediate and felsic batholiths rising from the subducting plate and intruding into the continental edge in the model. Three metamorphic facies occur in this situation. **Greenschist facies** is low-grade metamorphism and occurs farthest away from the heat and pressure. On its outer edge it merges with the unmetamorphosed country rock. **Amphibolite facies** is medium-grade metamorphism and occurs on the inside toward the source of heat and pressure. **Granulite facies** is high-grade metamorphism and occurs where the heat and pressure are greatest. Metamorphism past granulite facies in most cases results in the melting of the parent rock. This, of course, is an igneous process, and we have already explored that in a previous lab.

Regional metamorphism produces metamorphic rocks most people have heard of (such as slate, schist, and gneiss) and seen, whether they know it or not, because of their use as building stone. The most obvious feature of these rocks is their fabric, which is either foliated (leaflike layers of minerals or platy minerals with all their flat faces lined up) or mineral banding (alternating layers of different mineral composition) by which they can be recognized and identified. The presence of foliation and mineral banding tells us that two kinds of pressure were affecting the rock. Hydrostatic pressure resulting from burial which squeezes the rock in all directions equally and directed pressure which squeezes the rock from one primary direction. Directed pressure produces the foliation because it forces all the platy minerals to line up with their flat faces at right angles to the directed pressure.

Contact and Hydrothermal Metamorphism   Some rocks not directly caught up in tectonic processes play a more submissive role during metamorphism. These are crustal rocks of shallow to middle depths which are passively invaded by igneous intrusions and **contact metamorphosed** (see illustration below). In these cases the pressure is relatively low and undirected, but the temperature is quite high due to the heat escaping from the cooling magma. Different parent rocks have different responses. Shale is baked, something like a clay pot in a kiln, producing a rock called **hornfels**. Contact metamorphism produces rocks without foliation because there is no directed pressure during metamorphism to line the minerals up. The quartz grains in a quartz sandstone simply fuse together producing a quartzite.

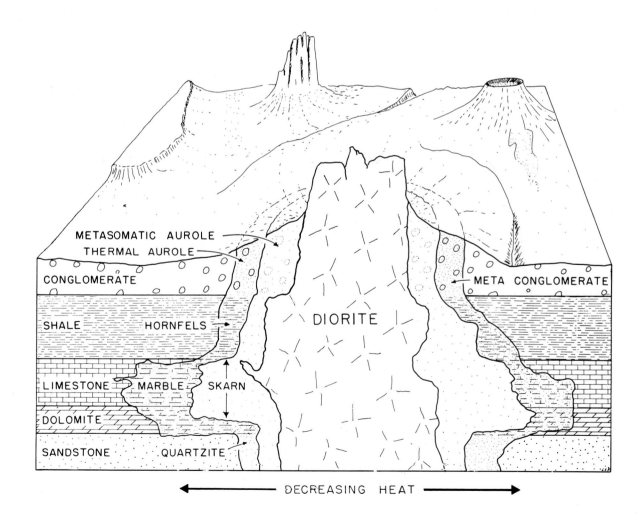

In addition to the escaping heat, chemical-rich fluids (metasomatic fluids) may also invade the country rock causing **hydrothermal metamorphism.** Hydrothermal metamorphic effects are very complex and interesting, but fall into three main categories.

1. Alteration of mafic igneous rocks into the minerals talc (rock = soapstone) and serpentine (rock = serpentinite).[1]
2. **Skarn** rocks formed when silica-rich fluids invade limestones and dolomites.
3. Mineral assemblages of calcite, galena, fluorite, and so on are deposited in rock cavities by fluids escaping a magma.

In the lab we deal only with hydrothermally altered mafic rocks. The other hydrothermal rocks are beyond the reach of our introduction to metamorphic rocks.

METAMORPHIC FACIES FOR THREE IDEAL PARENT ROCKS

One thing which can make metamorphic rocks difficult to study is that *any* kind of parent rock can undergo *any* kind of metamorphism, resulting in a very large number of combinations and permutations. Metamorphic petrologists deal with metamorphic rocks at this level, and we will not. For blueschist, eclogite, and contact metamorphism there will be only one representative metamorphic rock. For hydrothermal metamorphism there will be two representative rocks.

It is the regional metamorphic facies on which we concentrate, and even here the rock types to study can be reduced to a limited number. Even at this simple level, however, parent rock cannot be completely ignored. So we discuss the results of regional metamorphism for three very common parent rocks: mafic igneous rocks, felsic igneous rocks, and shaley or sandy-shaley sediments. The chart "Major Parent Rocks and their Metamorphic Equivalents" (next page) summarizes all this information.

Regional Metamorphism of Mafic Igneous Rocks   Mafic igneous rocks include basalt and gabbro. They are rich in pyroxene and calcic plagioclase, with occasional olivine. With the increasing temperature and pressure of metamorphism these minerals become unstable and begin to break down to form new minerals stable under the new conditions. At the low-grade temperatures and directed pressures, minerals such as chlorite and epidote form. Chlorite is a dark green platy mineral, like a mica. Because it is forming in a region of directed pressure, as the chlorite minerals grow all their flat faces line up forming a distinct foliation. Typically the chlorite crystals are large enough to see by eye. This rock is greenschist, but is sometimes also called a greenstone.[2]

If the temperature and pressure around the greenstone is raised to the amphibolite grade, the chlorite minerals are no longer stable and break down to form new minerals which are stable at the higher temperatures and pressures. These minerals are amphibole and plagioclase with a weak foliation forming the rock amphibolite, the most obvious representative of the amphibolite facies.

Raise the temperature and pressure even more, and the amphibole and plagioclase in the amphibolite become unstable, break down, and form new minerals stable under the new conditions. These minerals are quartz and feldspar associated with any of several other minerals typical of, or found only

---

[1] Mafic rocks are composed of minerals harder than glass. Talc and serpentine are both softer than your fingernail. It might be difficult to imagine that the soft metamorphic rocks soapstone and serpentinite can come from such a hard parent rock, but it is true.

[2] Do not confuse greenstone (greenschist) the rock with greenschist the facies. Greenstone is a field term applied to low-grade (greenschist facies) metamorphosed mafic igneous rock. Greenschist the facies implies the specific set of low-grade, regional metamorphic, temperature, and pressure conditions under which any rock may be metamorphosed.

in, mid -to high-grade metamorphic rocks, such as pyroxene, garnet, and kyanite. This rock is a granulite, the most typical representative of the granulite facies.

Thus, the result of regional metamorphism of a mafic igneous rock produces three rocks whose names are easy to remember because they are the same names as the facies: greenschist, amphibolite, granulite.

MAJOR PARENT ROCKS AND THEIR METAMORPHIC EQUIVALENTS

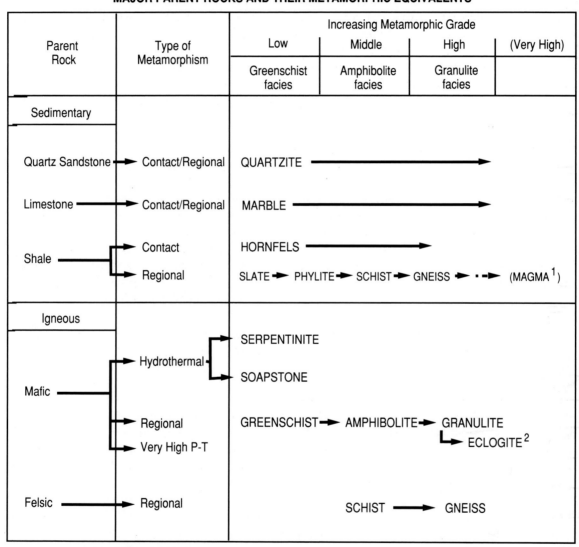

Parent Rock	Type of Metamorphism	Increasing Metamorphic Grade			
		Low	Middle	High	(Very High)
		Greenschist facies	Amphibolite facies	Granulite facies	
Sedimentary					
Quartz Sandstone	Contact/Regional	QUARTZITE ⟶			
Limestone	Contact/Regional	MARBLE ⟶			
Shale	Contact	HORNFELS ⟶			
	Regional	SLATE ➤ PHYLITE ➤ SCHIST ➤ GNEISS ➤ ··➤ (MAGMA[1])			
Igneous					
Mafic	Hydrothermal	SERPENTINITE / SOAPSTONE			
	Regional	GREENSCHIST ➤ AMPHIBOLITE ➤ GRANULITE			
	Very High P-T			⟶ ECLOGITE[2]	
Felsic	Regional		SCHIST ⟶ GNEISS		

Regional Metamorphism of Felsic Igneous Rocks   Felsic igneous rocks include rhyolite and granite. They are rich in quartz, orthoclase, and sodic plagioclase. These minerals are relatively stable during metamorphism and will tend to remain during regional metamorphism. There are two

[1] Beyond a certain temperature, and at low to intermediate pressures, gneiss will give way to granite. The transition zone from pure gneiss to pure granite is often a swirled mixture showing in one place gneissic mineral banding and in another place right next to it granitic texture. These rocks whose exact gneissic/granitic nature cannot be determined are called **migmatites.**

[2] **Eclogite** forms deep within the earth where pressures are high enough that melting of the rock does not occur.

changes resulting from metamorphism, however. First, the rocks develop a foliation or mineral banding and so become either a schist or a gneiss. Second, new minerals found only in metamorphic rocks, minerals such as garnet, kyanite, staurolite, and sillimanite, will grow in the rock. The typical rocks formed here are schists and gneisses (descriptions of schist and gneiss are below).

Regional Metamorphism of Sedimentary Rocks   A rock like a shale may seem to be a boring rock, but in fact it is not. A shale is compositionally very rich being composed of clay minerals (rich in silica, oxygen, potassium, and aluminum), as well as iron oxides, very fine mica grains, carbonate, and other impurities, Such a rich diversity of compounds and elements is capable of producing many new and different minerals and rocks during metamorphism.

As an example of how a sediment responds to regional metamorphism, we are going to take a shale, but it could be a sandy shale, or a silty sand, or a sandstone rich in lithic fragments and clay. We start the shale at the earth's surface and bury it progressively deeper, subjecting it to increasing temperature and pressure.

First   **The formation of slate**. The clay minerals which compose a shale are stable only near the earth's surface. If the clay is buried and subject to enough heat and pressure, it breaks down and forms a new mineral, **chlorite**, which is stable under the new conditions. Chlorite is a platy mineral (like mica) and normally dark green, although a slate may still retain its original sedimentary rock color and be gray, black, or red among other colors. The change in composition can be noted by tapping the rock on a hard surface: slate tends to ring rather than go "thunk" like a shale. As the chlorite crystals grow, the pressure arranges them so that all the flat surfaces line up forming **foliation**. This foliation causes the rock to break into perfectly flat, parallel layers. At first the chlorite crystals are too small to be seen by eye and overall the rock has a dull luster. The foliation always forms at right angles to the stress, so the rock may have two sets of layers running through it, one a trace of the former bedding, the other the foliation.

Thus a slate is different from a shale because it is composed of chlorite, has a ring (instead of a thunk) when struck, and breaks along perfect foliation layers. A shale may share some of these features, but not all of them.

Second   **Increase the heat and pressure and the slate turns into phyllite.** If the slate is further buried, and the heat and pressure increases more, the chlorite crystals continue to grow until they are large enough to produce a sheen on the surface which reflects light, but still not large enough to be seen without a microscope. Because there is still pressure the foliation remains, but now the rock tends to break along larger, more irregular planes. Also, the phyllite is usually the green color of the chlorite because most other minerals which can color a rock have broken down.

Third   **Increase the heat and pressure and the phyllite turns into a schist.** Beyond a certain temperature the chlorite minerals in the phyllite are no longer stable and begin to break down to form new minerals which are now stable under the new conditions. The minerals fall into two groups.

The first group of minerals are ones you already know, such as quartz, biotite, muscovite, and feldspar. In a schist these minerals are completely mixed together, but because they form under pressure, the mica minerals still all line up so a distinct foliation is still present. The foliation will keep you from confusing this rock with an igneous rock.

The second group of minerals are ones you have not yet studied because they are found only in metamorphic rocks, such as garnet, staurolite, kyanite, and sillimanite. One important thing to know about them is that they do not all form at once but in a specific order as the heat and pressure increases so they will not all be found in a schist together.

Thus, a schist is different from a slate and phyllite because chlorite is absent and different from an igneous rock of the same composition because there is foliation in the rock and exclusively metamorphic minerals may be present.

Fourth **Increase the heat and pressure and the schist turns into a gneiss.** By the gneiss stage, most of the new metamorphic minerals formed in a schist are no longer stable and break down. Now all that is left is quartz, feldspar, and mafic minerals such as biotite and amphibole. The foliation also disappears. Instead of these minerals being mixed as in a schist with all the mica minerals lined up, the quartz and feldspar separate into a band away from the biotite which is in bands of its own. The mineral banding in a gneiss is its most obvious feature.

Thus, the gneiss is not a schist because all the special index minerals are absent and because the biotite is separated into bands away from the quartz and feldspar. The gneiss is not an igneous rock, even though it can have the same mineral content, because of the mineral banding.

Fifth **Now the temperatures are getting very hot, and the quartz, feldspar, and biotite which form the gneiss are no longer stable and begin to break down.** But there are no other high-temperature minerals they can change into and so they melt. Now we have a magma of felsic composition. We are no longer dealing with metamorphic rocks but igneous rocks—and that is another subject.

These four rocks, slate, phyllite, schist, and gneiss, are typical of regional metamorphism and the metamorphic facies to which they belong is indicated in the diagram on page 106.

METAMORPHIC ROCK CLASSIFICATION

Because metamorphism is the alteration of *any* other rock and can take place by many combinations of heat, pressure, or hot fluids, in many different kinds of geological situations, the variety of metamorphic rocks is potentially large. Yet at a basic level we can divide metamorphic rocks into a limited number of kinds.

In the lab you will learn metamorphic rocks by first dividing them by their fabric, whether granular or foliated/mineral-banded. After that, rock identification will depend on identifying the minerals present in the rock. The classification will be relatively simple and straightforward.

What the classification will not do is lead naturally to an interpretation of metamorphic rocks. In this respect metamorphic rocks are different from igneous and sedimentary rocks, where the systematic use of texture and composition not only led to a classification, it also led to interpretation. Such a simple connection between classification and interpretation does not exist for metamorphic rocks. For example, both foliated and granular metamorphic rocks are produced by regional metamorphism, and granular rocks are found in regional, contact, and hydrothermal metamorphism. Yet there are systematic ways to understand metamorphic rock interpretation, and after you learn to recognize the basic rock types, we will concentrate on organizing them by interpretation.

Metamorphic Rocks

THIS LABORATORY ASSUMES YOU KNOW OR CAN DO THE FOLLOWING:

1. Distinguish among the conditions which produce **blueschist, regional, contact,** and **hydrothermal metamorphism.**

2. Define **rock fabric,** and distinguish between **granular, foliated,** and **mineral banded.**

3. Know what is meant by **metamorphic grade.**

4. Know why rocks like quartz sandstone and limestone produce only one kind of granular rocks, while rocks like shale produce a great diversity of metamorphic rocks.

   If you do not know the answers to these find out now. Read the Preliminary to Metamorphic Rocks, your notes, and your textbook, or ask your neighbor.

ORGANIZATION

There are four parts to this laboratory:

PART ONE—METAMORPHIC MINERALS   Many minerals in metamorphic rocks are the same ones learned for igneous rocks. But there are other minerals formed only during metamorphism, and you need to know them before identifying metamorphic rocks. These minerals have a special key (pages 112–113) in this laboratory.

PART TWO—METAMORPHIC ROCK IDENTIFICATION   Only a Level One classification of metamorphic rocks is presented here. This level will allow you to recognize and identify all the basic kinds of metamorphic rocks.

PART THREE—INTERPRETING METAMORPHIC ROCKS   This part explores where metamorphic rocks form and why. Use the fold out chart in the pocket at the back of the manual, *Kinds of Metamorphism.*

PART FOUR—METAMORPHIC ROCK CRITICAL REASONING PROBLEMS

─Part One─

METAMORPHIC
MINERALS

Because metamorphic rocks form under conditions unlike the conditions for igneous and sedimentary rocks, a variety of new minerals form in metamorphic rocks, minerals which are stable under the higher-temperature and -pressure conditions typical of metamorphism.

☐ (1) Get a tray of minerals from your instructor. Work together in groups of two or three.

☐ (2) Select a mineral from the tray and write its number (if available) on the chart on pages 111 and 114. If the specimens are not numbered, just arrange them in a row.

☐ (3) Determine the physical properties of the mineral and list them on the chart.

☐ (4) Use the "Key to the Identification of MetamorphicMinerals" on pages 112–113 and identify the mineral.

☐ (5) After you have identified two or three minerals, ask your instructor to check them before identifying the rest to make sure you are working correctly.

☐ (6) When all the minerals are identified and you feel confident you know them:
   A. Scramble all the minerals. Can you still identify them?
   B. Exchange your tray for another and look at different specimens of the same minerals. Can you still identify them?

PHYSICAL PROPERTIES

SPECIMEN
NUMBER

MINERAL
NAME

| HARDNESS | LUSTER | CLEAVAGE |
| STREAK | COLOR | OTHER |

| HARDNESS | LUSTER | CLEAVAGE |
| STREAK | COLOR | OTHER |

| HARDNESS | LUSTER | CLEAVAGE |
| STREAK | COLOR | OTHER |

| HARDNESS | LUSTER | CLEAVAGE |
| STREAK | COLOR | OTHER |

| HARDNESS | LUSTER | CLEAVAGE |
| STREAK | COLOR | OTHER |

| HARDNESS | LUSTER | CLEAVAGE |
| STREAK | COLOR | OTHER |

| HARDNESS | LUSTER | CLEAVAGE |
| STREAK | COLOR | OTHER |

| HARDNESS | LUSTER | CLEAVAGE |
| STREAK | COLOR | OTHER |

| HARDNESS | LUSTER | CLEAVAGE |
| STREAK | COLOR | OTHER |

| HARDNESS | LUSTER | CLEAVAGE |
| STREAK | COLOR | OTHER |

KEY TO THE IDENTIFICATION OF METAMORPHIC MINERALS

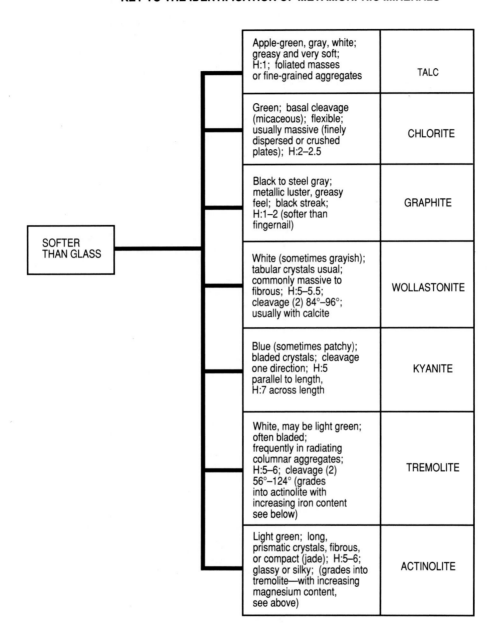

SOFTER THAN GLASS		
	Apple-green, gray, white; greasy and very soft; H:1; foliated masses or fine-grained aggregates	TALC
	Green; basal cleavage (micaceous); flexible; usually massive (finely dispersed or crushed plates); H:2–2.5	CHLORITE
	Black to steel gray; metallic luster, greasy feel; black streak; H:1–2 (softer than fingernail)	GRAPHITE
	White (sometimes grayish); tabular crystals usual; commonly massive to fibrous; H:5–5.5; cleavage (2) 84°–96°; usually with calcite	WOLLASTONITE
	Blue (sometimes patchy); bladed crystals; cleavage one direction; H:5 parallel to length, H:7 across length	KYANITE
	White, may be light green; often bladed; frequently in radiating columnar aggregates; H:5–6; cleavage (2) 56°–124° (grades into actinolite with increasing iron content see below)	TREMOLITE
	Light green; long, prismatic crystals, fibrous, or compact (jade); H:5–6; glassy or silky; (grades into tremolite—with increasing magnesium content, see above)	ACTINOLITE

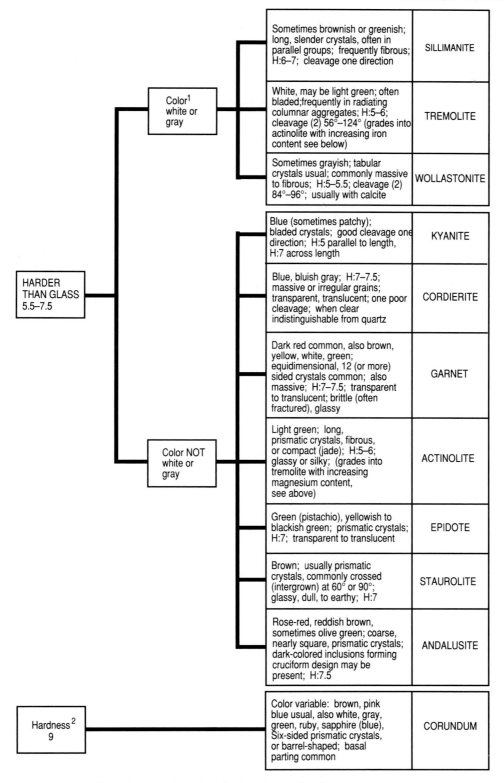

		Sometimes brownish or greenish; long, slender crystals, often in parallel groups; frequently fibrous; H:6–7; cleavage one direction	SILLIMANITE
HARDER THAN GLASS 5.5–7.5	Color[1] white or gray	White, may be light green; often bladed; frequently in radiating columnar aggregates; H:5–6; cleavage (2) 56°–124° (grades into actinolite with increasing iron content see below)	TREMOLITE
		Sometimes grayish; tabular crystals usual; commonly massive to fibrous; H:5–5.5; cleavage (2) 84°–96°; usually with calcite	WOLLASTONITE
		Blue (sometimes patchy); bladed crystals; good cleavage one direction; H:5 parallel to length, H:7 across length	KYANITE
		Blue, bluish gray; H:7–7.5; massive or irregular grains; transparent, translucent; one poor cleavage; when clear indistinguishable from quartz	CORDIERITE
		Dark red common, also brown, yellow, white, green; equidimensional, 12 (or more) sided crystals common; also massive; H:7–7.5; transparent to translucent; brittle (often fractured), glassy	GARNET
	Color NOT white or gray	Light green; long, prismatic crystals, fibrous, or compact (jade); H:5–6; glassy or silky; (grades into tremolite with increasing magnesium content, see above)	ACTINOLITE
		Green (pistachio), yellowish to blackish green; prismatic crystals; H:7; transparent to translucent	EPIDOTE
		Brown; usually prismatic crystals, commonly crossed (intergrown) at 60° or 90°; glassy, dull, to earthy; H:7	STAUROLITE
		Rose-red, reddish brown, sometimes olive green; coarse, nearly square, prismatic crystals; dark-colored inclusions forming cruciform design may be present; H:7.5	ANDALUSITE
Hardness[2] 9		Color variable: brown, pink blue usual, also white, gray, green, ruby, sapphire (blue). Six-sided prismatic crystals, or barrel-shaped; basal parting common	CORUNDUM

[1]These three minerals resemble each other and may be difficult to distinguish in hand specimen.

[2]Often lab specimens are badly weathered. The weathering product may not even scratch glass! Look for a fresh shiny surface when testing for hardness.

PHYSICAL PROPERTIES

SPECIMEN
NUMBER

MINERAL
NAME

HARDNESS	LUSTER	CLEAVAGE
STREAK	COLOR	OTHER

HARDNESS	LUSTER	CLEAVAGE
STREAK	COLOR	OTHER

HARDNESS	LUSTER	CLEAVAGE
STREAK	COLOR	OTHER

HARDNESS	LUSTER	CLEAVAGE
STREAK	COLOR	OTHER

HARDNESS	LUSTER	CLEAVAGE
STREAK	COLOR	OTHER

HARDNESS	LUSTER	CLEAVAGE
STREAK	COLOR	OTHER

HARDNESS	LUSTER	CLEAVAGE
STREAK	COLOR	OTHER

HARDNESS	LUSTER	CLEAVAGE
STREAK	COLOR	OTHER

HARDNESS	LUSTER	CLEAVAGE
STREAK	COLOR	OTHER

HARDNESS	LUSTER	CLEAVAGE
STREAK	COLOR	OTHER

HARDNESS	LUSTER	CLEAVAGE
STREAK	COLOR	OTHER

─ *Part Two* ─────────────────────

METAMORPHIC ROCK
IDENTIFICATION ────────

At this level you will do two things. First, divide your rocks into groups based on fabric and color. Second, identify the major kinds of metamorphic rocks.

☐   (1)   Get a tray of metamorphic rocks from your instructor. Work together in pairs.

☐   (2)   SEPARATE ROCKS BY FABRIC   Divide your rocks into two piles by fabric: first pile, foliated rocks—that is, rocks which show a distinct layering or mineral banding; second pile, granular rocks—that is, rocks in which all the mineral grains are equidimensional in shape and of about the same size. This means there is no preferred orientation of minerals in the rock; it looks just about the same in all directions.

> **OBSERVE:** Be cautious about color banding or streaking which may look superficially like foliation. Foliation/mineral banding is directly related to the arrangement of minerals in the rock. If the color bands or streaks do not correspond to the arrangement of minerals, then the rock is granular.

☐   (3)   Ask your instructor to check your piles based on fabric.

☐   (4)   Using the "Keys to the Identification of Metamorphic Rocks," pages 118–119, identify your metamorphic rocks. Use the data sheets on pages 116–117 and 120–122 to record your observations for each rock.

☐   (5)   Once all the rock are identified, ask your instructor to check your identifications.

> **OBSERVE:** The metamorphic minerals you identified in Part One are frequently just small pieces of rock. You should not arbitrarily place mineral and rock specimens in these different categories. Go back and examine your mineral specimens, but see if they can also be classified as rocks as well as minerals.

DECISION TREES FOR IDENTIFYING
METAMORPHIC ROCKS

In the laboratory on minerals we contrasted the brute force method for learning minerals and rocks ("I'll just stare at the specimens so long maybe I'll recognize them on the exam") with the hypothesis-test method (see pages 13–14). The hypothesis-test method helps you to understand why rocks are different from each other and to develop a strategy for confidently identifying unknown specimens.

☐ (1) If they are available, pick out and lay in front of you the following sets of metamorphic rocks.
   A. Marble, quartzite, hornfels
   B. Phyllite, greenschist, serpentinite
   C. Schist, gneiss, quartzite
   D. Hornfels, serpentinite, soapstone

☐ (2) On a separate piece of paper draw a decision tree like those you did in the *Minerals* lab, page 14. You may not use color.
   When finished compare your decision trees with ours at the back of the chapter. Please do not look at our decision trees before you draw your own. If you are unable to draw a decision tree which works, then you should get help.

☐ (3) Now pick out any other metamorphic rocks which confuse you. Ask, "What distinctive characteristics will separate these rocks?" Develop a hypothesis-test strategy which *you* can use to separate and identify the specimens.

DATA SHEETS FOR METAMORPHIC ROCKS

ROCK DESCRIPTION   SPECIMEN NUMBER   ROCK NAME

TEXTURE	FABRIC	HARDNESS
☐ Fine	☐ Granular	☐ > Glass
☐ Visible	☐ Foliated	☐ < Glass
MINERALS	☐ Mineral banding	☐ < Fingernail
	ROCK COLOR	OTHER
	ACID REACTION	

_____ [ ]

TEXTURE	FABRIC	HARDNESS
☐ Fine	☐ Granular	☐ > Glass
☐ Visible	☐ Foliated	☐ < Glass
MINERALS	☐ Mineral banding	☐ < Fingernail
	ROCK COLOR	OTHER
	ACID REACTION	

_____ [ ]

TEXTURE	FABRIC	HARDNESS
☐ Fine	☐ Granular	☐ > Glass
☐ Visible	☐ Foliated	☐ < Glass
MINERALS	☐ Mineral banding	☐ < Fingernail
	ROCK COLOR	OTHER
	ACID REACTION	

_____ [ ]

DATA SHEETS FOR METAMORPHIC ROCKS

ROCK DESCRIPTION

SPECIMEN NUMBER

ROCK NAME

Specimen 1

TEXTURE	FABRIC	HARDNESS
☐ Fine	☐ Granular	☐ > Glass
☐ Visible	☐ Foliated	☐ < Glass
MINERALS	☐ Mineral banding	☐ < Fingernail
	ROCK COLOR	OTHER
	ACID REACTION	

Specimen 2

TEXTURE	FABRIC	HARDNESS
☐ Fine	☐ Granular	☐ > Glass
☐ Visible	☐ Foliated	☐ < Glass
MINERALS	☐ Mineral banding	☐ < Fingernail
	ROCK COLOR	OTHER
	ACID REACTION	

Specimen 3

TEXTURE	FABRIC	HARDNESS
☐ Fine	☐ Granular	☐ > Glass
☐ Visible	☐ Foliated	☐ < Glass
MINERALS	☐ Mineral banding	☐ < Fingernail
	ROCK COLOR	OTHER
	ACID REACTION	

Specimen 4

TEXTURE	FABRIC	HARDNESS
☐ Fine	☐ Granular	☐ > Glass
☐ Visible	☐ Foliated	☐ < Glass
MINERALS	☐ Mineral banding	☐ < Fingernail
	ROCK COLOR	OTHER
	ACID REACTION	

Specimen 5

TEXTURE	FABRIC	HARDNESS
☐ Fine	☐ Granular	☐ > Glass
☐ Visible	☐ Foliated	☐ < Glass
MINERALS	☐ Mineral banding	☐ < Fingernail
	ROCK COLOR	OTHER
	ACID REACTION	

GRANULAR METAMORPHIC ROCKS
KEY TO IDENTIFICATION

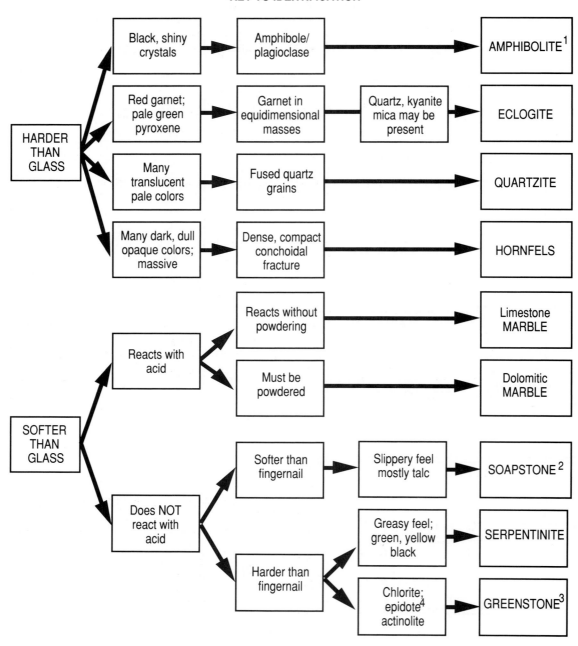

[1]Amphibolite is usually foliated but some specimens may appear granular.
[2]May be weakly foliated.
[3]Greenstone is usually well foliated but massive varieties exist.
[4]Epidote is a pale green. Often it is finely disseminated in the rock so individual crystals cannot be seen. Look for pale green patches within the darker green chlorite matrix.

FOLIATED AND BANDED METAMORPHIC ROCKS
KEY TO IDENTIFICATION

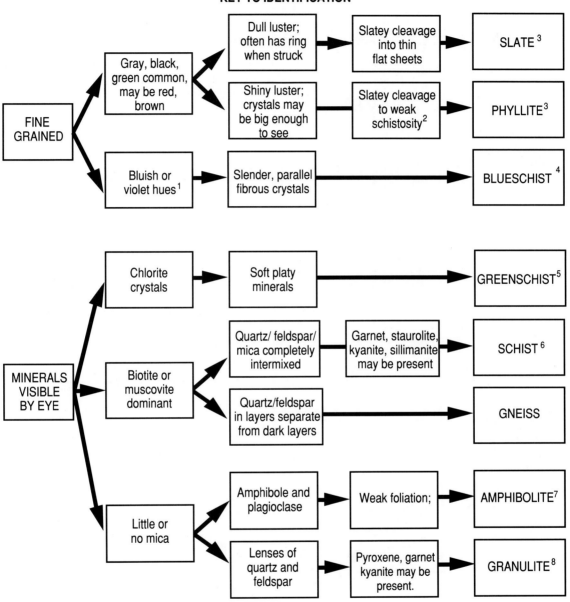

[1] Under fluorescent light bluish hues may not be easy to detect. On the outcrop in full daylight, rock is usually distinctly blue in color.

[2] Schistosity = coarse-grained foliation. Phyllites frequently have an undulatory surface and are not flat like slates and shales.

[3] (Shale), slate, and phyllite completely intergrade with each other. Distinctions may be difficult. Ask your instructor.

[4] Blueschist is also called glaucophane schist.

[5] Greenschist may superficially look like slate/phyllite but has moderately developed schistosity (see note 2).

[6] Rock may be called *garnet schist*, or *kyanite schist*, or *garnet-kyanite schist*, and so on, depending on accessory minerals present.

[7] Amphibolite may be granular in appearance.

[8] Granulite may be crudely gneissic or granular in appearance.

DATA SHEETS FOR METAMORPHIC ROCKS

ROCK DESCRIPTION   SPECIMEN NUMBER   ROCK NAME

TEXTURE	FABRIC	HARDNESS
☐ Fine	☐ Granular	☐ > Glass
☐ Visible	☐ Foliated	☐ < Glass
MINERALS	☐ Mineral banding	☐ < Fingernail
	ROCK COLOR	OTHER
	ACID REACTION	

TEXTURE	FABRIC	HARDNESS
☐ Fine	☐ Granular	☐ > Glass
☐ Visible	☐ Foliated	☐ < Glass
MINERALS	☐ Mineral banding	☐ < Fingernail
	ROCK COLOR	OTHER
	ACID REACTION	

TEXTURE	FABRIC	HARDNESS
☐ Fine	☐ Granular	☐ > Glass
☐ Visible	☐ Foliated	☐ < Glass
MINERALS	☐ Mineral banding	☐ < Fingernail
	ROCK COLOR	OTHER
	ACID REACTION	

TEXTURE	FABRIC	HARDNESS
☐ Fine	☐ Granular	☐ > Glass
☐ Visible	☐ Foliated	☐ < Glass
MINERALS	☐ Mineral banding	☐ < Fingernail
	ROCK COLOR	OTHER
	ACID REACTION	

TEXTURE	FABRIC	HARDNESS
☐ Fine	☐ Granular	☐ > Glass
☐ Visible	☐ Foliated	☐ < Glass
MINERALS	☐ Mineral banding	☐ < Fingernail
	ROCK COLOR	OTHER
	ACID REACTION	

DATA SHEETS FOR METAMORPHIC ROCKS

ROCK DESCRIPTION

SPECIMEN NUMBER

ROCK NAME

TEXTURE	FABRIC	HARDNESS
☐ Fine	☐ Granular	☐ > Glass
☐ Visible	☐ Foliated	☐ < Glass
MINERALS	☐ Mineral banding	☐ < Fingernail
	ROCK COLOR	OTHER
	ACID REACTION	

TEXTURE	FABRIC	HARDNESS
☐ Fine	☐ Granular	☐ > Glass
☐ Visible	☐ Foliated	☐ < Glass
MINERALS	☐ Mineral banding	☐ < Fingernail
	ROCK COLOR	OTHER
	ACID REACTION	

TEXTURE	FABRIC	HARDNESS
☐ Fine	☐ Granular	☐ > Glass
☐ Visible	☐ Foliated	☐ < Glass
MINERALS	☐ Mineral banding	☐ < Fingernail
	ROCK COLOR	OTHER
	ACID REACTION	

TEXTURE	FABRIC	HARDNESS
☐ Fine	☐ Granular	☐ > Glass
☐ Visible	☐ Foliated	☐ < Glass
MINERALS	☐ Mineral banding	☐ < Fingernail
	ROCK COLOR	OTHER
	ACID REACTION	

TEXTURE	FABRIC	HARDNESS
☐ Fine	☐ Granular	☐ > Glass
☐ Visible	☐ Foliated	☐ < Glass
MINERALS	☐ Mineral banding	☐ < Fingernail
	ROCK COLOR	OTHER
	ACID REACTION	

DATA SHEETS FOR METAMORPHIC ROCKS

ROCK DESCRIPTION			SPECIMEN NUMBER	ROCK NAME

Block 1

TEXTURE	FABRIC	HARDNESS
☐ Fine	☐ Granular	☐ > Glass
☐ Visible	☐ Foliated	☐ < Glass
MINERALS	☐ Mineral banding	☐ < Fingernail
	ROCK COLOR	OTHER
	ACID REACTION	

Block 2

TEXTURE	FABRIC	HARDNESS
☐ Fine	☐ Granular	☐ > Glass
☐ Visible	☐ Foliated	☐ < Glass
MINERALS	☐ Mineral banding	☐ < Fingernail
	ROCK COLOR	OTHER
	ACID REACTION	

Block 3

TEXTURE	FABRIC	HARDNESS
☐ Fine	☐ Granular	☐ > Glass
☐ Visible	☐ Foliated	☐ < Glass
MINERALS	☐ Mineral banding	☐ < Fingernail
	ROCK COLOR	OTHER
	ACID REACTION	

Block 4

TEXTURE	FABRIC	HARDNESS
☐ Fine	☐ Granular	☐ > Glass
☐ Visible	☐ Foliated	☐ < Glass
MINERALS	☐ Mineral banding	☐ < Fingernail
	ROCK COLOR	OTHER
	ACID REACTION	

Block 5

TEXTURE	FABRIC	HARDNESS
☐ Fine	☐ Granular	☐ > Glass
☐ Visible	☐ Foliated	☐ < Glass
MINERALS	☐ Mineral banding	☐ < Fingernail
	ROCK COLOR	OTHER
	ACID REACTION	

Part Three

INTERPRETING METAMORPHIC ROCKS

In the study of igneous and sedimentary rocks, the use of texture and composition led naturally to both identification and interpretation. With metamorphic rocks, the criteria used for identification do not always lead naturally to interpretations. Metamorphism occurs in too many ways and has too many parent rocks for such simple relationships to exist. Our first task of interpretation, therefore, is to learn to associate each metamorphic rock or sequence of metamorphic rocks with the kinds of metamorphism which produce them.

☐   (1)   There are two things you should learn here:
- A. To develop a picture, in your mind's eye, of the chart "Kinds of Metamorphism."
- B. To be able to accurately place on that chart, in your mind's eye, various kinds of metamorphic rocks, and explain why they must be found in that particular place.

                   When you are done you should be able to point to any part of the chart, identify the types of metamorphism, and say what kinds of metamorphic rock would be found there.

☐   (2)   You need to have two things: a tray of metamorphic rocks and the chart in the pocket at the back of the manual titled "Kinds of Metamorphism."

☐   (3)   You are to arrange your metamorphic rocks on the chart in their right space.

                   You should not expect to have samples of every kind of metamorphism. By arranging all your metamorphic rocks on the chart, you will not only know what rocks you have seen and identified, but what kinds of metamorphic rocks you have not yet seen.

☐   (4)   On page 125 is a chart similar to the large chart titled "Kinds of Metamorphism." Note that in the field for each type of metamorphism are small boxes. Write in the appropriate boxes the names of all the rock types you have on your large chart. Write in a different color the names of all the rocks you are missing. Below is a list of the kinds of rocks which you might have in your collection and should be looking for.

> **OBSERVE:** The metamorphic sequences on the chart on page 125 begin with the three ideal parent rocks discussed in the Preliminary, but different parent rocks may produce similar metamorphic rocks. Check the chart on page 106 in the Preliminary, "Major Parent Rocks and Their Metamorphic Equivalents."

Contact Metamorphism—high temperature, low pressure
     Rock = Hornfels
Hydrothermal Metamorphism—hot fluids
     Rocks = Serpentinite, Soapstone

Regional Metamorphism—low to high temperature, mid to high pressure. Regional metamorphism includes

FACIES: Greenschist ➡ Amphibolite ➡ Granulite ➡ (Magma)

ROCKS: Slate/Phyllite | Schist | Gneiss | (Magma)

ROCKS: Greenschist | Amphibolite | Granulite | (Magma)
(Greenstone)

Blueschist Metamorphism—low temperature, very high pressure
Rock = Blueschist

OBSERVE: The hydrothermal rocks serpentinite and soapstone form from the alteration of mafic igneous rocks. Other kinds of hydrothermal rocks exist, such as **skarn** rocks formed when silica-rich fluids invade limestones and dolomites, and mineral assemblages of calcite, galena, fluorite, and so on are deposited in rock cavities. We deal only with hydrothermally altered mafic rocks.

OBSERVE: This cannot be said enough. There is a difference between greenschist, amphibolite, and granulite **facies** and greenschist, amphibolite, and granulite **rocks.** The facies are associations of rocks which formed under the same temperature and pressure conditions. The specific rocks greenschist (sometimes greenstone), amphibolite, and granulite are rocks formed from the metamorphism of mafic parent rocks under the conditions of each metamorphic facies. If the parent rock were a shale subject to greenschist facies metamorphism, the new rocks would be slate and phyllite; if amphibolite facies metamorphism, the new rock would be schist; and if granulite facies metamorphism, the new rock would be gneiss. If you study the chart "Major Parent Rocks And Their Metamorphic Equivalents" on page 106, you can see that any particular facies of metamorphism can result in many different specific metamorphic rocks depending on what the parent rock was.

OBSERVE: The magma at the end of some sequences above, of course, is in the realm of igneous rocks. To make your large chart complete, you will need to get an igneous rock.

METAMORPHIC INDEX MINERALS

The intensity of metamorphism is measured with specific minerals, some of which form under very narrow temperature/pressure conditions. Through extensive laboratory experiments, a sequence of index minerals has been found. An **index mineral** is one which forms and is stable under a very specific set of temperature and pressure conditions. These conditions are so well known that if we were to walk across a metamorphic countryside until we first found a particular index mineral, we could say with great confidence, "At this spot, at the time this rock formed, the temperature and pressure conditions were x and y."

There are six index minerals which define six metamorphic **zones** from the lowest-grade metamorphism (greenschist facies) to the highest-grade metamorphism (granulite facies). In order, these minerals are chlorite, biotite, garnet, staurolite, kyanite, and sillimanite. In the figure at the top of page 126 you can see that a relationship exists between the sequence of metamorphic rock (slate, phyllite, schist, and gneiss) and the six metamorphic zones.

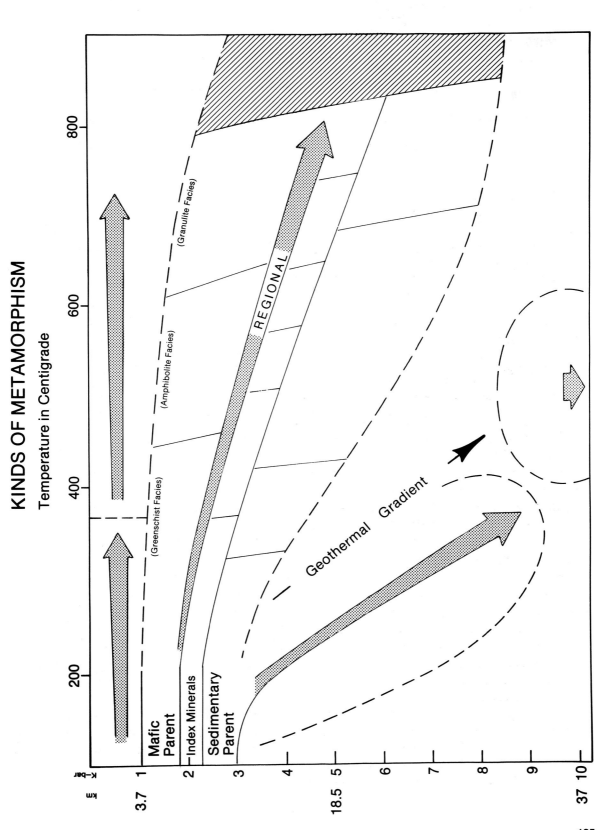

KINDS OF METAMORPHISM
Temperature in Centigrade

Mafic Parent
Index Minerals
Sedimentary Parent

(Greenschist Facies)
(Amphibolite Facies)
(Granulite Facies)

REGIONAL

Geothermal Gradient

K-bar
km
3.7
18.5
37

Chlorite $\Rightarrow$ Biotite $\Rightarrow$ Garnet $\Rightarrow$ Staurolite $\Rightarrow$ Kyanite $\Rightarrow$ Sillimanite

Greenschist Facies	Amphibolite Facies	Granulite Facies
Slate Phyllite	Schist	Gneiss

☐   (5)   We now want you to rearrange your regional metamorphic rocks by index minerals. It is unlikely you have enough rocks to have all the six index minerals represented. So, get from your tray of metamorphic minerals examples of the index minerals to supplement the rocks.

☐   (6)   Write the name of the index minerals you have in the appropriate boxes on the chart "Kinds of Metamorphism" on page 125. Write in a different color the names of the index minerals you are missing.

LEFTOVER ROCKS

You should now have some rocks left over, including perhaps quartzite, marble, and anthracite coal. They are not included in the chart "Kinds of Metamorphism" not because they are not common or important. These particular rocks are just not typical of a single type of metamorphism.

Part Four

CRITICAL REASONING PROBLEMS

☐   (1)   Write answers to the following critical reasoning problems about metamorphic rocks following these instructions:

     A.   For each problem there is a statement followed by several possible answers. For each possible answer you must indicate whether you REJECT it or ACCEPT it.

     B.   Logically and factually, in *writing*, explain your acceptance or rejection of *each* and *every* answer. There is NO CREDIT for a "right" answer, only for the analysis.

     C.   For each problem there is one ACCEPT and two REJECTS. Some answers will differ by subtle distinctions, but you must find the best one to ACCEPT and explain why the other answers are less correct.

     D.   You may discuss the problem with classmates, but when you write your analysis, it must be your own thinking, in your own writing.

☐   (2)   For each problem find examples of each of the rocks and minerals and lay them out so you can examine them while you formulate and write your reasons for acceptance or rejection.

Problem Number One

ACCEPT one, REJECT two. A limestone metamorphoses into:

_____ 1.   A foliated rock during regional metamorphism.

_____ 2.   A foliated rock during contact metamorphism.

_____ 3.   A granular rock during contact and regional metamorphism.

Problem Number Two

ACCEPT one, REJECT two. Contact metamorphism:

_____ 1.   Forms foliated rocks under some circumstances.

_____ 2.   Forms principally granular rocks.

_____ 3.   Forms chlorite-rich rocks.

Problem Number Three

ACCEPT one, REJECT two. Only regional metamorphism:

_____ 1.   Produces a schist.

_____ 2.   Produces a granular rock.

_____ 3.   Produces a hornfels.

Problem Number Four

ACCEPT one, REJECT two. A shale metamorphoses into:

_____ 1.   A granular rock during regional metamorphism.

_____ 2.   A foliated rock during contact metamorphism.

_____ 3.   A granular rock during contact metamorphism.

Problem Number Five

ACCEPT one, REJECT two. Granular metamorphic rocks:

_____ 1.   Form only during regional metamorphism.

_____ 2.   Form only during contact metamorphism.

_____ 3.   Form in both regional and contact metamorphism.

Problem Number Six

ACCEPT one, REJECT two. Very-high-pressure and low-temperature metamorphic rocks are most likely to be associated with:

_____ 1.   A volcanic pipe.

_____ 2.   Rocks surrounding a batholith.

_____ 3.   A subduction zone.

Problem Number Seven

ACCEPT one, REJECT two. During hydrothermal metamorphism a basalt turns into:

_____ 1.   A foliated rock composed of chlorite.

_____ 2.   A granular rock composed of chlorite.

_____ 3.   A rock composed of serpentine.

Problem Number Eight

ACCEPT one, REJECT two. A foliated metamorphic rock composed of chlorite most likely resulted from:

_____ 1.   Regional metamorphism of a shale.

_____ 2.   Contact metamorphism of a shale.

_____ 3.   Hydrothermal metamorphism of a basalt.

Problem Number Nine

ACCEPT one, REJECT two. A phyllite metamorphosed one step farther becomes:

_____ 1.   A granulite.

_____ 2.   A soapstone.

_____ 3.   A schist.

Problem Number Ten

ACCEPT one, REJECT two. High-temperature and medium-to -high-pressure regional metamorphic rocks derived from a shale are more likely to be:

_____ 1.   Chlorite rich.

_____ 2.   Serpentine rich.

_____ 3.   Quartz rich.

DECISION TREES FOR DISTINGUISHING AMONG METAMORPHIC ROCKS LISTED ON PAGE 109

Decision Tree for Marble, Quartzite, and Hornfels

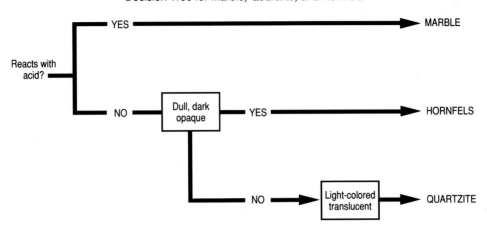

Decision Tree for Phyllite, Greenschist, and Serpentinite

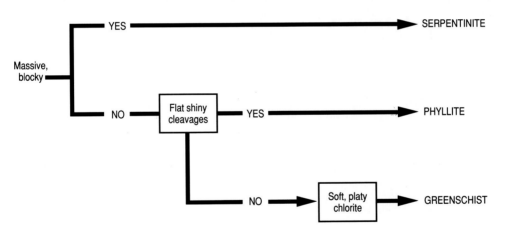

Decision Tree for Schist, Gneiss, and Quartzite

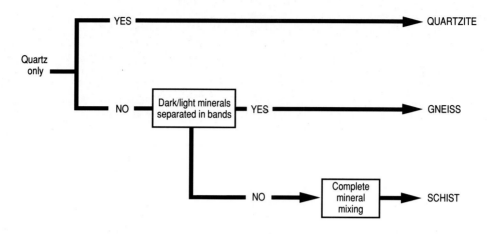

Decision Tree for Hornfels, Serpentinite, and Soapstone

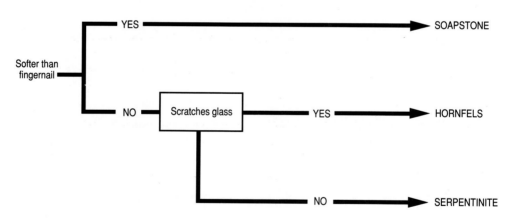

Preliminary to Earth Deformation and Geologic History

PURPOSE

Igneous, sedimentary, and metamorphic rocks probably seemed complicated and difficult when you first began to study them. Yet we found that by understanding just a few rules and processes, they could be recognized, classified, and interpreted with little difficulty. Having a framework of rules and processes makes the study of rocks much easier. Similarly, with a few more rules and processes, we can begin to interpret earth deformation and geologic history. This Preliminary introduces these new rules and processes, plus some new terminology needed to interpret earth history.

By the time you enter the laboratory you should:

1. Be able to recognize **anticlines** and **synclines** and know how they form.

2. Be able to recognize **normal, reverse, lateral, vertical, thrust,** and **oblique** faults and know how they form.

3. Know the **laws of original horizontality, superposition,** and **cross-cutting relationships** and how these are used to interpret geologic history.

4. Know the three kinds of **unconformities** and how these are used to interpret geologic history.

INTRODUCTION

One of the things which makes the earth unique in the solar system is that it has remained geologically very active since its origin about 4.5 billion years ago, while virtually all the other terrestrial bodies (planets and moons) in the solar system are geologically dead, or have extremely limited geologic activity very different from that on earth. We do not know why the earth alone is so active, but the geological activity has been critically important to you and me. It is quite likely that the evolution of life and the evolution of the earth are closely coupled, that each occurred as it did because of the presence of the other. Without the continuous geologic activity, the evolution of life would undoubtedly have been very different, and the human species may never have evolved. Also, without geologic activity the continents would never have formed and grown to such large size, and virtually none of the mineral resources our civilization depends on would have formed.

Many geologists study earth deformation and geologic history just out of curiosity, and entire courses are devoted to exploring various aspects of it. But these studies also have many practical applications which affect our lives everyday. Understanding the earth's deformation and geologic history is an important step to finding mineral resources, to understanding geologic hazards, and to deciding the best places to build highways, buildings, and landfills. Important decisions influencing all our lives are made nearly everyday based on this knowledge.

For you to interpret earth history, you must first know the specific conditions under which individual igneous, sedimentary, and metamorphic rocks form. From the labs you have done so far, you have a good understanding of individual rock formation and interpretation. Given any igneous, sedimentary, or metamorphic rock you should be able to describe the conditions which existed at the time and place the rock formed. This is the first part of the study of earth history.

The second part of the study of earth history is figuring out how individual rocks got to their present location. Because the earth has been ceaselessly active for billions of years, most rocks have been moved, sometimes several times, from their original location. Many rocks now side by side were originally tens of thousands of miles apart. It is also not uncommon to find side by side rocks which formed under very different conditions, for example, deep-forming granite in contact with surface-forming sedimentary rocks. It is also not uncommon to find rocks folded, broken, and mixed together like a jigsaw puzzle. Much research must still be done to understand the earth's deformation and geologic history, and what you will learn in the laboratory is the foundation on which it is built.

SEDIMENTARY ROCKS AND ORIGINAL HORIZONTALITY

Sedimentary rocks come in great variety and form in many different environments. Yet one thing they all have in common is that when they are deposited they are deposited flat lying or nearly flat lying. This is referred to as the **law of original horizontality.** Even when there are initial irregularities in the depositional surface, the low areas tend to be filled in first, and a horizontal surface is soon realized, as shown in the figure below. There are exceptions to this law, but the original horizontality is general enough that it is safe to assume that, unless there is evidence to the contrary, a sedimentary rock was deposited horizontally.

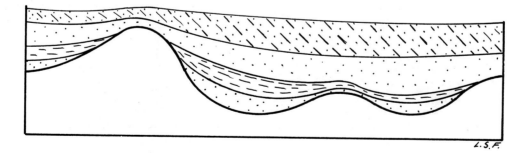

If we accept original horizontality, then it logically follows that when we find sedimentary rocks which are not horizontal, they have been deformed. There are many places on the earth where the sedimentary rocks show little or no deformation. The central part of North America from Canada to the Gulf of Mexico is an example. Most of the rocks found there are flat-lying. In other parts of the earth it may be difficult to find any rocks which are still horizontal. This is true in most mountain regions such as the Appalachian and Rocky mountain systems.

Sedimentary rocks which are not horizontal are said to be **dipping**. In some cases the rocks are not only dipping, they have been turned upside down; these are said to be **overturned beds.**

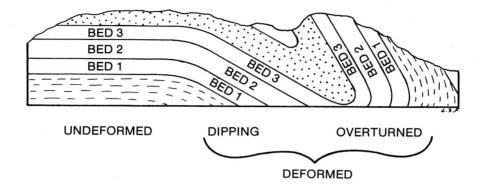

UNDEFORMED   DIPPING   OVERTURNED

DEFORMED

ANTICLINES AND SYNCLINES

For sedimentary beds to be deformed some outside force has to be applied. Two basic kinds of forces are **compression**, which squeezes the rocks together, and **tension,** which pulls the rocks apart. Rocks are not strong under tension. Instead of deforming they usually break, that is, are faulted. Faulting frequently results in blocks of the earth sliding and tilting, and any sedimentary rocks on the block are naturally tilted and are no longer horizontal. Destroying original horizontality by faulting is common (see next section on faulting), but deformation under compression is our major focus here.

During compression sedimentary beds are folded. It may seem unreasonable that a hard, brittle rock could fold, especially in complex folds where the rocks seem to have behaved like taffy, stretching, twisting, and distorting in complex ways. Folding does not occur at the surface, however, but deep inside the earth where the rocks are under great pressure. Under pressure, rocks behave like a soft plastic and fold easily. A logical consequence of this is that all the folds we now see at the surface must have been deeply buried in the past.

Upfolds, that is, folds which form an arch, are called **anticlines.** Downfolds, that is, folds which form a trough, are called **synclines.** Anticlines and synclines are the two most common kinds of folds, and although they may be seen alone, they commonly form together in a series of alternating anticlines and synclines.

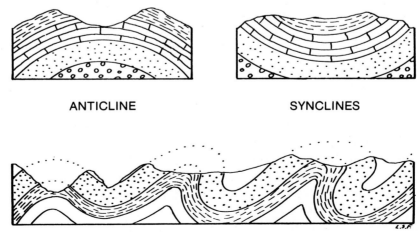

ANTICLINE   SYNCLINES

ALTERNATING ANTICLINES AND SYNCLINES

At first it may seem that anticlines would form hills and synclines form valleys, but this may not be the case at all. What controls **topography** (lay of the land) is the relative resistance of rocks to weathering and erosion. Sandstones generally weather more slowly than shales, for instance. In regions where these two are exposed sandstones form ridges while shales are found in valleys. Thus, hills and valleys do not correspond directly with anticlines and synclines, as the illustration on page 133 shows.

Anticlines and synclines do not extend forever but must eventually descend below the surface, and these are called **plunging folds.** A plunging anticline is illustrated below.

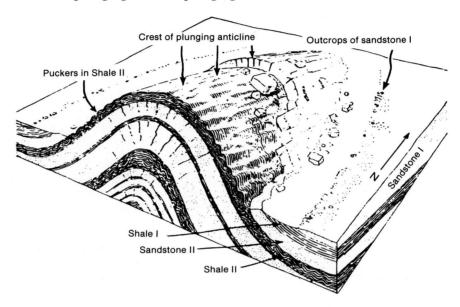

FAULTS: BREAKS IN THE EARTH'S CRUST

Sometimes when a force is applied to a rock it deforms plastically and folds. At other times the rocks are too brittle to fold and then they break, forming a crack along which no movement has occurred; these cracks are called **joints.** If the rocks on the two sides of the joint slide past each other, however, the joint surface becomes a **fault.**

Three kinds of forces cause faulting:

1. *Compressive forces* are squeezing forces which shorten the earth's crust. Shortening may occur by the formation of anticlines and synclines if the rocks behave plastically. But if the rocks are brittle, they break and form compressional faults such as **thrust** and **reverse** faults.

2. *Tension forces* are stretching forces which tend to lengthen the earth's crust. No major folding occurs under tension, but the rocks easily break and are pulled apart allowing blocks of the earth to slide down past each other along faults such as in **normal** and **vertical** faults.

3. *Shear forces* are those which cause two blocks to slide past each other in opposite directions without up and down movement and result in **lateral** or **strike-slip** faults.

KINDS OF FAULTS

Faults in the earth's crust are very common and range from microscopic in size to those hundreds of miles long. Many technical names are given to all the different kinds of faults that have been found, but for our purposes the great variety can be classified into three categories based on the break's orientation: dipping breaks, vertical breaks, and nearly horizontal breaks.

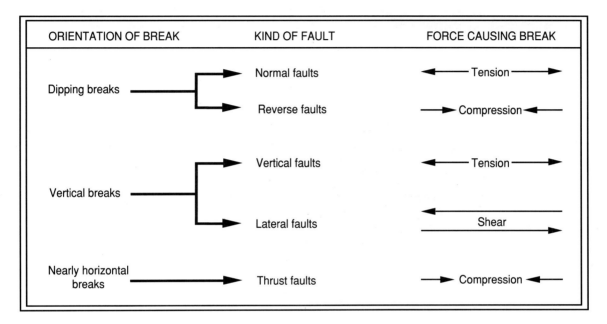

ORIENTATION OF BREAK	KIND OF FAULT	FORCE CAUSING BREAK

RECOGNIZING DIFFERENT KINDS OF FAULTS

Within the three categories of breaks in the earth are five different kinds of faults identified by the relative direction the blocks on either side of the fault have moved: normal, reverse, vertical, lateral, and thrust (see illustrations on page 136)

VERTICAL FAULTS   are ones in which one block has moved *up* along a vertical break while the other has moved down. In the block diagram, it is easy to see which block has gone up and which down, but in the earth it is not always obvious. In naming the fault it does not matter which block has moved; it is the relative motion along a vertical break that is important.

LATERAL FAULTS   occur when two blocks move past each other because of shear forces. Some of the largest faults in the world are lateral faults, such as the San Andreas fault cutting through California.[1] One of the things we are very interested in with lateral faults like the San Andreas is the relative motion of the two sides of the fault. Motion along lateral faults is either *left lateral* or *right lateral* and is easy to determine. For example, imagine that you are standing on the east side of the San Andreas fault looking west, out toward the Pacific Ocean. The part of California on the

---

[1] The San Andreas fault is also a **transform fault.** A transform fault is a special case of a lateral fault; that is, not all lateral faults are transform faults, but all transform faults are lateral faults. A transform fault forms one of the boundaries between crustal plates, major divisions of the earth's crust.

ILLUSTRATIONS OF THE MAJOR KINDS OF FAULTS

DIPPING BREAKS

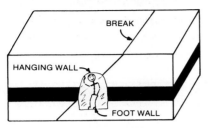

UNFAULTED
(WITH MINE SHAFT)

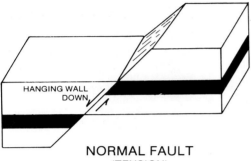

NORMAL FAULT
(TENSION)

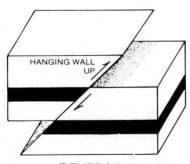

REVERSE FAULT
(COMPRESSION)

VERTICAL BREAKS

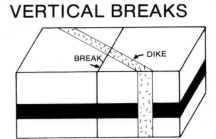

UNFAULTED

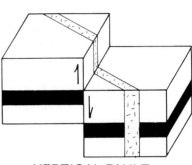

VERTICAL FAULT
(TENSION)

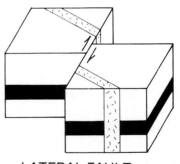

LATERAL FAULT
(SHEAR)

NEARLY HORIZONTAL BREAKS

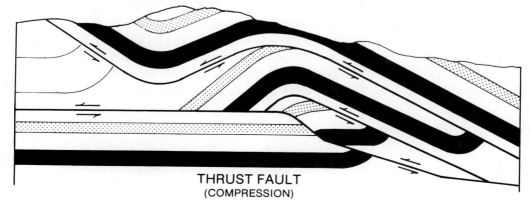

THRUST FAULT
(COMPRESSION)

opposite(west) side of the fault from where you are standing moves to your right, so the San Andreas is a right lateral fault. But what if you are standing on the west side of the fault looking east toward the rest of North America and the Atlantic? It does not matter, for now the rest of North America and the Atlantic on the opposite (east) side of the fault from where you are standing is also moving to your right, making it, still, a right lateral fault. All lateral faults are this way; the relative motion of the blocks is the same regardless of which side of the fault you are standing on.

NORMAL and REVERSE FAULTS   are sometimes a problem for beginners to recognize and identify, but there are two easy ways to remember them. First the names. Normal faults are called normal because as the earth is pulled apart under tension, a gap is created and it is *normal* for one block of the earth to slide down under gravity into the gap. Reverse faults are called reverse because during compression one block is forced to slide up over the other in the *reverse* direction of gravity.

Second, is a classic way to recognize normal and reverse faults. Faults are weaknesses in the earth crust along which important minerals are often deposited. Many mines have been dug along faults and in the mine shafts miners would stand on a side of the fault called the **foot wall**, which is the lower wall (see upper left figure on page 136). The top of the shaft then hangs over their head and is called the **hanging wall**. The hanging wall is always the higher, or upper, wall even if the fault plain is nearly vertical. This terminology leads to the following rules:

RULE ONE   If the *hanging wall* has moved *up*, the fault is *reverse*.

RULE TWO   If the *hanging wall* has moved *down*, the fault is *normal*.

If you find a fault confusing, just draw a little mine shaft on it and determine which is the hanging wall. Its direction of movement determines the type of fault.

THRUST FAULTS   occur when relatively thin sheets of rock[1] slide up, across, and over adjacent parts of the earth's surface. Movement along thrust faults is typically miles to tens of miles. Sometimes a single thrust fault occurs with one thrust sheet, but more commonly several, or many, thrust faults occur stacked on top of each other.

Notice in the cross section at the bottom of page 136 the several thrust faults stacked on top of each other. Notice how some thrust faults are almost flat (horizontal) while other thrust faults are folded. Notice how thrust faults can split apart at one place and join back together at another. Notice the anticlines and synclines in the cross section; these folds often result when one thrust sheet slides up and over a lower thrust sheet.

Because thrust faults are so large, only very small fragments of any one fault can be seen at one time, which sometimes makes them difficult to interpret. For example, look at the uppermost thrust fault, the folded one, in the cross section on page 136. Notice how, near its center (left to right), the thrust fault dips down towards the left. If you saw the thrust fault only at that place, it would look like a normal fault because it appears as if a hanging wall has moved down. Only by reconstructing the whole fault can its thrust nature be discerned. Because of difficulties like these, and thrust fault size, geologists have only recently achieved clear understanding of thrust faults.

Thrust faults are some of the most common and important faults in the world and are found in such mountain ranges as the Appalachians, parts of the Rockies, the Alps in Europe, and the Himalayas.

---

[1]"Thin" is a relative term; thrust sheets can be thousands of feet thick and cover tens of thousands of square miles.

UNTANGLING EARTH HISTORY

Geology is historical. The rocks at any one place on the earth are rarely if ever the same age. One of the first things geologists must do when entering a new area is to unravel the geologic history of the region, that is, the order in which the various rocks formed, folded, faulted, and eroded. Unraveling the history of the earth is a study in its own right, but nearly all branches of geology make contributions to it. Everything you have learned so far are necessary tools and knowledge needed to interpret earth history.

SOME SIMPLE LOGIC Since a historical study involves placing events in their proper sequence, there must be some rules for deciding the order in which events have occurred. Some of these rules you already know, but they all involve simple logic.

1. **Law of Original Horizontality.** Sedimentary rocks are originally deposited horizontally. Therefore, sedimentary rocks which are not now horizontal have been through a period of deformation.

2. **Law of Superposition.** In a sequence of sedimentary rocks those on the bottom were deposited first and are the oldest while those on top were deposited last and are the youngest.

3. **Law of Cross-Cutting Relationships.** When a rock body (such as dike) cuts across another body of rock, *the one that cuts across must be younger*. This rule applies no matter what is cutting what. A fault which cuts a rock must be younger than the rock. A fault which cuts another fault must be younger than the faults it cuts.

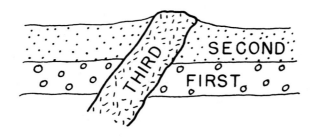

THE UNCONFORMITY

An unconformity is a gap in the historical rock record where no direct or visible evidence exists to indicate what events occurred during that time. There are three kinds of unconformities:

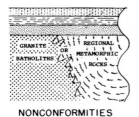

NONCONFORMITIES

ANGULAR
UNCONFORMITIES

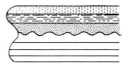

DISCONFORMITY

NONCOMFORMITY   Igneous batholithic rocks and/or regional metamorphic rocks are overlain by sedimentary rocks. Since the batholithic/regional metamorphic rocks form only deep in the earth and sedimentary rocks form only at the earth's surface, a very long period of erosion must have occurred to bring the batholithic/regional metamorphic rocks to the surface.

ANGULAR UNCONFORMITY   Previous sedimentary rocks have been deformed, so they are no longer horizontal, and eroded so that individual rock units are no longer continuous but truncated (cut off). At some later date newer sedimentary rocks are deposited on the deformed and truncated older rocks. The angular contact between these two sets of rocks implies a period of deformation and erosion.

DISCONFORMITY   Following formation of a series of sedimentary rocks, either (1) deposition ceases, leaving a gap in the record, or (2) erosion occurs, removing some of the originally deposited rock. Later deposition places new, younger rocks on top of the older rocks. In either case no deformation occurs, no angular contacts exist, and the rocks appear to have been deposited continuously without break. But time is missing due to the lack of deposition or erosion. Usually drawn as a wavy line, representing erosion, on illustrations.

INTERPRETING GEOLOGIC HISTORY

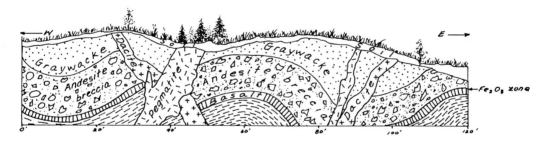

The illustration above is a cross section of a part of the earth where several different kinds of rocks are present either stacked on top of each other or in cross-cutting relationships. Our task is to begin with the oldest rock in the region and arrange the entire sequence of rocks and events (e.g., deformation events, faulting events, erosion events) in the order in which they occurred. This is the skill you will learn in the laboratory.

Below are listed in order the sequence of geologic events preserved in the cross section above. Study the cross section and our interpretation of it. Notice how the rules of simple logic are used, and the way the evidence in the rocks is used to make interpretations. Notice also that we include as individual steps in the list of events not only the formation of each rock unit but also periods of deformation and erosion (unconformities).

It is possible for interpretations to become very sophisticated, but learning how to make basic interpretations is not at all difficult. All that is required is knowing how the common rocks form and systematically applying the rules of simple logic to them. By now you have enough knowledge to do this.

EVENT	EVIDENCE AND/OR INTERPRETATION
1. The rock units were formed in the following order.	1. Law of superposition
A. Deposition of shale	A. Shale is on the bottom; therefore, it was deposited first.
B. Basaltic unit	B. This unit could be a sill or a lava flow. It has to be a lava flow and not a sill because of the weathered Fe_2O_3 zone. If it had been a sill, it could not have formed at this time in the sequence of events.
C. Deposition of andesite breccia	C. Despite the composition (andesite—an igneous rock) this is a sedimentary rock (breccia) formed from weathering, erosion, and accumulation near a volcanic highland source.
D. Deposition of graywacke	D. Classically formed by rapid sedimentation of partially weathered rock debris in a deep ocean trench.
2. Episode of deformation	2. Law of original horizontality. Since these rocks are no longer horizontal, they must have been deformed. Folding indicates that the force was compression.
3. Normal faulting and intrusion of dacite (an igneous rock)	3. Normal faulting because the beds have been offset and the hanging wall has moved down. Dacite intrusion along joints and faults. Normal faulting indicates that the force was tension. The dacite intrusions along joints and faults have not been folded, which makes these events postdeformation.
4. Intrusion of pegmatite	4. Law of cross-cutting relationships. The **pegmatite** cuts across all prior events; therefore, it is younger than all prior events.
5. Uplift and erosion forming unconformity at base of soil	5. The beds and igneous intrusions are truncated (cut off).
6. Development of soil and vegetation cover	6. Soil and vegetation blankets all prior events.

Earth Deformation and Geologic History

All the rocks which form on the earth are the result of, or the response to, deformation in the earth's crust. Conversely, each rock contains the information needed to reconstruct the conditions which existed at the time it formed. Therefore, interpreting earth history is a two-step process: first, interpreting each individual rock, and second, interpreting the sequence of deformational events through which all the rocks in a region have formed. Doing this requires that you apply all the skills and knowledge you have learned so far and integrate them with new skills you will learn in this laboratory.

THIS LABORATORY ASSUMES YOU KNOW OR CAN DO THE FOLLOWING:

1. Recognize **anticlines** and **synclines** and the forces that produce them.

2. Recognize **normal, reverse, lateral, vertical, thrust,** and **oblique** faults, and know how they form.

3. Know the **laws of original horizontality, superposition**, and **cross-cutting relationships,** and how these are used to interpret geologic history.

4. Know the three kinds of **unconformities** and how these are used to interpret geologic history.

   If you do not know the answers to these find out now. Read the Preliminary to Earth Deformation and Geologic History, your notes, your textbook, or ask your neighbor.

There are three parts to this laboratory:

PART ONE—SIMPLE EXERCISES IN STRUCTURAL GEOLOGY   These simple exercises will help you become familiar with perspective drawings used in structural geology and the solution of simple structural geology problems.

PART TWO—CRITICAL REASONING PROBLEMS

PART THREE—INTERPRETING GEOLOGIC HISTORY FROM GEOLOGIC CROSS SECTIONS   Seven geologic cross sections ranging from simple to complex for practicing rock interpretation and the interpretation of geologic history.

Part One

SIMPLE EXERCISES IN STRUCTURAL GEOLOGY

DEFINING THE ATTITUDE OF ROCK UNITS

The **attitude** of a rock unit is its orientation: flat-lying, standing on end, or tilted in some direction. In areas where sedimentary beds are lying horizontal, it is easy to describe their attitude: they are flat. But in areas where rocks have been deformed in complex ways, beds may be tilted in many ways, and rocks only a few hundred feet apart may be tilted in opposite directions.

An additional problem is that all the changes in attitude must be described on a flat piece of paper, a map, and it is impractical to try to describe in writing all the different attitudes which may be present. To deal with this, geologists have devised symbols which quickly describe all the different attitudes a bed can have. You need to know the most common symbols.

STRIKE AND DIP The most important and common symbol for describing a bed's attitude is the strike and dip symbol.

STRIKE is a line formed by the intersection of a horizontal plane with a plane dipping at an angle. It is described as the compass direction toward which the line points. For example, we would say, "That line is striking north," which means the dipping plane is striking north also.

DIP is the angle a plane is tilted and ranges from 90° to anything above 0°. Dip is always measured where the angle is largest, and this occurs only at right angles to strike.

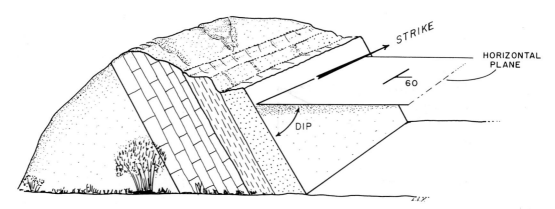

The drawing above illustrates strike and dip, but if you are having trouble visualizing it, try this. Prop a book on your desk at an angle. Take a sheet of paper and, holding it perfectly horizontal, move it until one edge of the paper rests against the tilted book cover. With some chalk, draw a line where the paper and the book cover intersect—that line is strike. Notice that the strike line is horizontal, and it points in a compass direction. If the strike line is pointing toward the northeast, we would then say that the plane of the book cover is striking northeast.

Hold the piece of paper back along the strike line and notice that there is an angle between the paper and the tilted book cover —that angle is dip. Use the chalk to draw a line at right angles to, and in the center of, the strike line *in the direction the book cover is dipping*. Notice that if you draw the dip line any way other than right angles to strike, the angle is less that true dip. If this is unclear, try drawing a dip line parallel to strike and decide what that angle is.

OBSERVE: You may have difficulty visualizing strike and dip at first and may find that you have to return to study it a number of times. Don't worry, this is normal. It takes practice to learn to visualize three-dimensional things drawn in two dimensions. If you are having trouble, ask your instructor for help. There are many ways to explain strike and dip, and one of them will work for you.

The standard symbol for strike and dip consists of two lines, a long line for strike, and a shorter line drawn at right angles to the center of the strike line pointing in the direction of dip. Sometimes the angle of dip is written in at the end of the dip line.

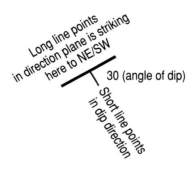

Long line points in direction plane is striking here to NE/SW

30 (angle of dip)

Short line points in dip direction

HORIZONTAL AND VERTICAL BEDS The standard strike and dip symbol can be applied to all planes, including dikes and faults, with two exceptions. Horizontal beds have no dip and, therefore, no strike either. The symbol for horizontal beds is a circle with a cross in it.

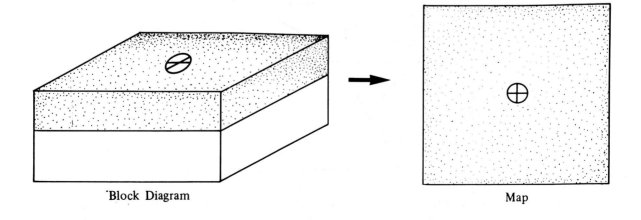

Block Diagram

Map

Vertical beds do have a strike but also dip 90° in two directions. The symbol is a strike line with two dip lines pointing in opposite directions.

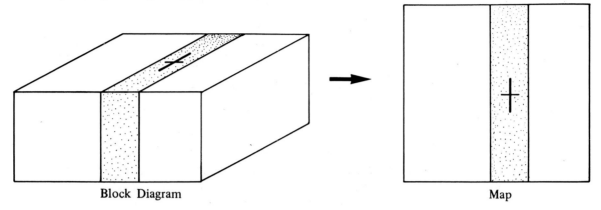

Block Diagram   Map

GEOLOGIC MAPS, BLOCK DIAGRAMS, AND CROSS SECTIONS

A **geologic map** shows the distribution of rocks on the ground and the contacts between different rocks as if seen from the air. Geologic maps are pretty straightforward since we can see the rocks at the surface, but we also want to know what happens to the rock units below ground. To show what happens to rocks below ground we draw geologic **cross sections** and **block diagrams** (see illustration below). Cross sections and block diagrams must be interpretations because rarely can we get below the surface to see where the rocks really go, and the most important skill you must learn here and in the lab on geologic maps is to make these interpretations.

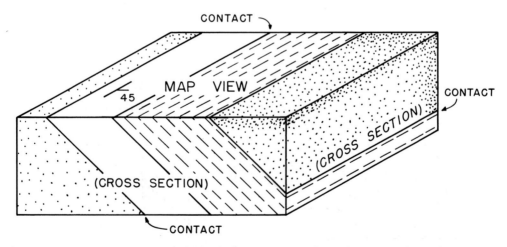

Geologic maps, cross sections, and block diagrams are all used commonly and interchangeably, and you must be careful to distinguish among them since the rock units drawn on each may not look that much different. The exercises on pages 145-147 are designed to help you visualize rock units as drawn on maps, cross sections, and block diagrams.

☐   (1)   On the next few pages are a series of incomplete geologic maps (right) and geologic block diagrams (left). Complete the missing portions of the map and/or block diagram.

☐   (2)   On all map views add the appropriate symbol for strike and dip, vertical beds, or horizontal beds. Be sure all sides of the block diagrams are finished.

☐   (3)   Ask your instructor to check your interpretations.

OBSERVE: Always complete the geologic map portion of the block diagram first since it is the map which determines where bed contacts go on the sides of the block diagram.

OBSERVE: Block diagrams are perspective drawings, and although they create an image which looks normal, some people find them difficult to draw at first because perspective drawings distort the image we see in our mind's eye, which is normally straight on. *A hint:* Notice that many contacts you need to draw run parallel to the sides of the block diagram. When you draw your contact lines, just make sure they also run parallel to the edge of the block.

OBSERVE: Be careful to draw angles accurately. If you do not have a protractor, angles can be easily estimated. Draw a 90° right angle; bisecting it gives 45°; bisecting that gives 22 $\frac{1}{2}$°. Note, however, that on the oblique sides of the block diagram angles are distorted.

OBSERVE: INSTRUCTIONS FOR BLOCK MODELS.[1] If you are having difficulty visualizing how to complete the perspective structure problems on pages 145-147, the four pages of 3–D blocks at the end of the chapter, which you can make and then rotate and view at any angle, may help. Cut the diagrams out, fold along the dashed lines, and glue or tape the tabs on the inside of the block. The small diagram to the left shows which parts of the flat sheets form which parts of the 3–D blocks. The blocks provided are similar to selected problems from pages 145-147 but not identical. One blank 3-D block is provided on p. 177 for fashioning other problems you may have difficulty visualizing in two dimensions.

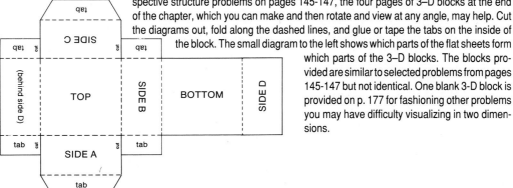

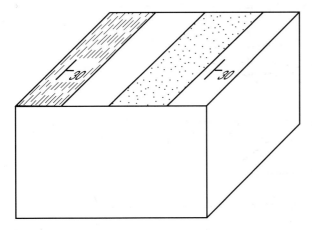
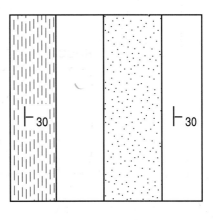

[1]Thanks and acknowledgments to Skip Watts, now at Radford University, who first devised the prototypes for these blocks many years ago.

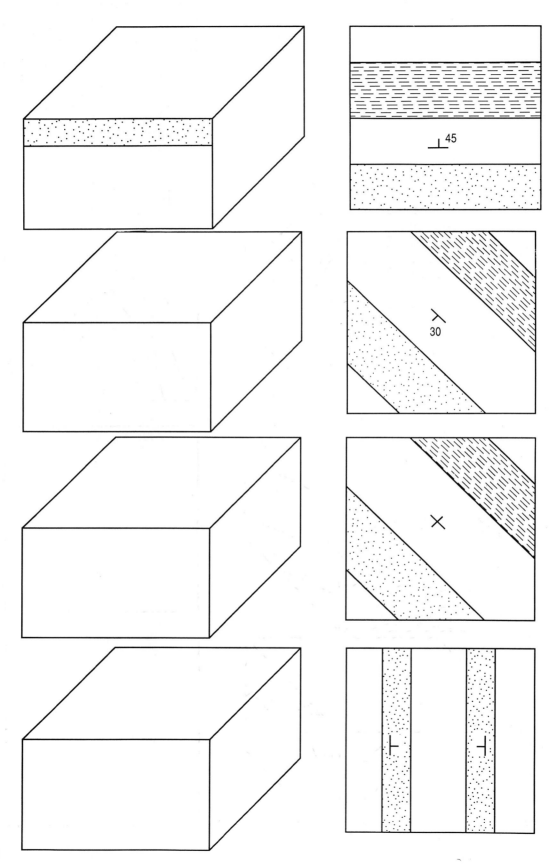

LEARNING TO UNDERSTAND FAULTS

Faults break and separate rock units, and before the geologic history of a region can be written, we must recognize how the rocks have moved along each fault, identify the kind(s) of fault(s) present, decide in what order the faults occurred if there is more than one, and be able to imagine how the area would have looked before the faulting occurred. In areas where the deformation is complex, this can be quite a challenge. The key to interpreting complex deformation is to be able to recognize the individual faults and analyze them one by one. The simple exercises that follow are designed for you to learn how to do this.

☐   (1)   Study each of the block diagrams on page 149 and complete the pattern of the rocks as they would appear after faulting.

            A.   Choose your own amount of movement (**displacement**) along the fault, but don't fault the marker bed out of sight.

            B.   Assume that each block, after faulting, has its dimensions restored so that the original size and shape of the block is present.

☐   (2)   Ask your instructor to check your interpretations.

MULTIPLE FAULTS   More than one fault may break the rocks in an area. Then you must not only be able to identify each kind of fault, you must also be able to decide the order in which the faults occurred.

☐   (1)   Determine the sequence of faults in the block diagram below. Label their ages as 1, 2, and 3. Name each of the faults and explain why each comes in the order it does.

☐   (2)   Ask your instructor to check your interpretations.

☐   (3)   There is a general rule which can be formulated for solving all questions of which came first in a series of cross-cutting relationships. Circle the choice in the rule below which is correct; the statement will become your general rule.

RULE   In a series of cross-cutting relationships, the unit which has been broken the (most, least) number of times is the oldest. The unit which has not been broken at all is the (oldest, youngest).

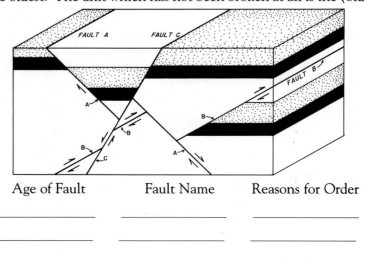

	Age of Fault	Fault Name	Reasons for Order
Fault A	_____	_____	_____
Fault B	_____	_____	_____
Fault C	_____	_____	_____

BEFORE FAULTING   AFTER FAULTING   FAULT NAME

_____

_____

_____

_____

Part Two

CRITICAL REASONING PROBLEMS

☐ (1) Provide answers to the following critical reasoning problems following these instructions:

   A. For each problem there is a structural situation presented, followed by several possible solutions. For each possible solution you must indicate whether you REJECT it or ACCEPT it.

   B. Logically and factually explain your acceptance or rejection of *each* and *every* answer. There is NO CREDIT for a "right" answer, only for the analysis.

   C. For each problem there is one ACCEPT and two REJECTS. Some answers will differ by subtle distinctions, but you must find the best one to ACCEPT and explain why the other answers are not as accurate.

   D. You may discuss the problem with classmates but when you write your analysis it must be your own thinking, in your own writing.

☐ (2) To describe localities on the block diagrams and maps, use compass directions as shown in Problems Number Two and Three. For example, you might say for Problem Number Two, "The southern block has moved up along a reverse fault, and after erosion the position of the dike on the southern block will appear to have shifted to the west."

Problem Number One

ACCEPT one, REJECT two. When three faults are all in cross-cutting relationships with each other, the one of middle age is:

_____ 1. Cut by the younger fault but not the older.

_____ 2. Cut by both the other faults.

_____ 3. Cut by the older fault but not the younger.

SPECIAL INSTRUCTIONS FOR PROBLEMS TWO THROUGH ELEVEN

Each of the following problems is presented as a geologic map or block diagram which is going to undergo faulting. You must imagine what happens during the faulting and decide which of the three solutions is correct.

Assume that after faulting the higher block (called the "upthrown" block) is eroded down to the level of the other block. Note that your analyses may be done as drawings of block diagrams or cross sections which demonstrate that a solution does or does not work.

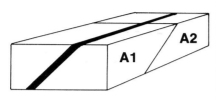

OBSERVE: INSTRUCTIONS FOR FAULTING MODELS. If you are having difficulty visualizing how to solve the critical reasoning problems, there are two sets of 3–D blocks at the end of the chapter that you can make which will illustrate some of the problems. Cut the diagrams out, fold along the dashed lines, and glue or tape the tabs on the inside of the block. The blocks fold together simply, but you will know if the block is folded together right if the dikes continuously wrap around the block. If you orient the block as in the drawing to the left, with the letters showing on the side and the fault dipping toward you, then the model will correspond to the following critical reasoning problems.

Problem #2   Blocks A1-A2   Problem #9   Blocks B1-B2
Problem #5   Blocks A1-A2   Problem #12   Blocks B1-B2
Problem #8   Blocks A1-A2

The dikes drawn on the 3–D blocks may not be located exactly as they are in the problems, but they will be similar enough to help solve the problem. If you orient the blocks in ways other than shown in the drawing above, then other kinds of faults and dike orientations can be studied.

Problem Number Two

ACCEPT one, REJECT two. Imagine that the block diagram on the left undergoes reverse faulting followed by erosion of the up-thrown block. The result would be:

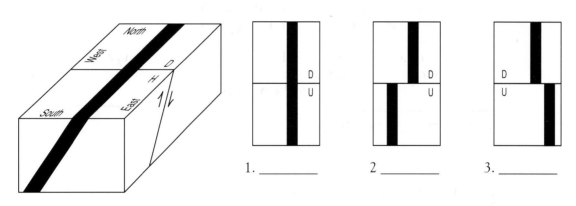

1. _____   2. _____   3. _____

Problem Number Three

ACCEPT one, REJECT two. Imagine that the structure on the map on the left is vertically faulted, followed by erosion of the up-thrown block. The result would be:

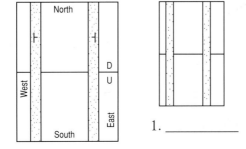
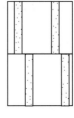
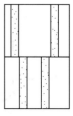

1. _____   2. _____   3. _____

Problem Number Four

ACCEPT one, REJECT two. Based on the evidence on the map in Problem Number Five, the fault is:

_____ 1.   Normal.

_____ 2.   Vertical.

_____ 3.   Reverse.

Problem Number Five

ACCEPT one, REJECT two. Imagine the structure on the map is faulted followed by erosion. The result would be:

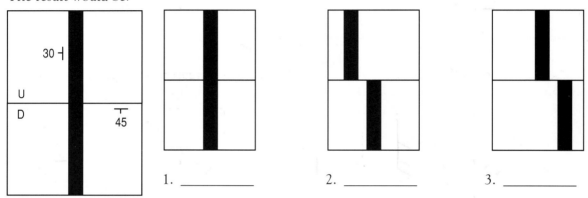

1. _____

2. _____

3. _____

Problem Number Six

ACCEPT one, REJECT two. Imagine the structure on the map is vertically faulted, followed by erosion. The result would be:

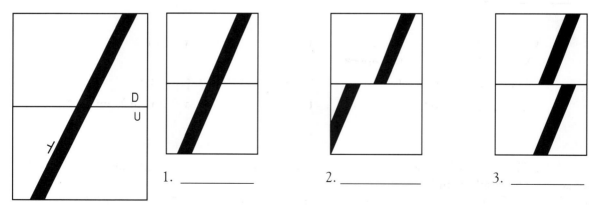

1. _____

2. _____

3. _____

Problem Number Seven

ACCEPT one, REJECT one. If the fault in Problem Number Six were normal instead of vertical, then the dike on the north block would:

1.   _____   move the same as in Problem Number Six.

2.   _____   move in the opposite direction of that in Problem Number Six.

Problem Number Eight

ACCEPT one, REJECT two. Imagine the block diagram undergoes normal faulting followed by erosion. The result would be:

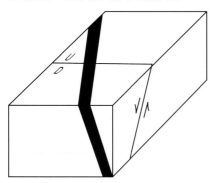

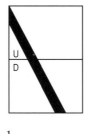

1._____

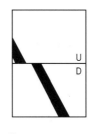

2._____

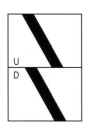

3._____

Problem Number Nine

ACCEPT one, REJECT two. Imagine the block diagram undergoes reverse faulting followed by erosion. The result would be:

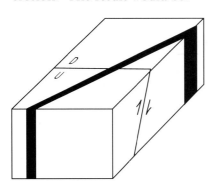

1._____

2._____

3._____

Problem Number Ten

ACCEPT one, REJECT one. If the fault in Problem Number Nine were normal instead of reverse, the dike on the north block would appear:

1.   _____   to move the same way as in Problem Number Nine.

2.   _____   to move in the opposite direction of that in Problem Number Nine.

Problem Number Eleven

ACCEPT one, REJECT two. Imagine the structure on the map is normally faulted, followed by erosion. The result would be:

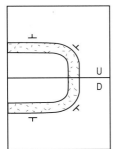

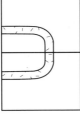

1._____

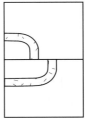

2._____

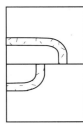

3._____

Problem Number Twelve

ACCEPT one, REJECT two. Imagine the block diagram undergoes normal faulting followed by erosion. The result would be:

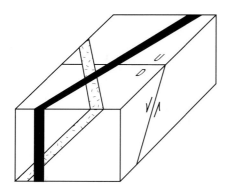 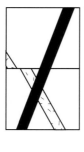 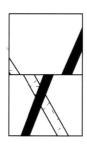 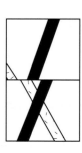

1._____   2._____   3._____

Part Three

INTERPRETING GEOLOGIC HISTORY FROM GEOLOGIC CROSS SECTIONS

In this section you will learn to interpret geologic cross sections. There are two ways to approach this, one haphazard and disorganized, the other deliberate and systematic. At first it takes more time and effort to learn to interpret cross sections deliberately and systematically, but...

> **OBSERVE:** We can virtually guarantee you that if you try to interpret geologic cross sections in a haphazard and unorganized way you will never get them right. One of the most important things you can learn here is that the solution of scientific problems must be done in as deliberate and systematic a manner as possible.

The object of these exercises is not just to get them done, but to develop the habit of developing a strategy for solving complex problems in a deliberate and systematic manner. A deliberate and systematic analysis means that you first take the cross section apart and understand all the individual pieces and then synthesize all the individual pieces into a whole interpretation. A well-analyzed cross section has notes written all over it. The example below shows one way to analyze a cross section. We have done the analysis in a series of steps which you can use as you learn to do your analyses. In time, you will develop your own strategy of analysis.

- ☐ (1) Work together in groups of two or three. It helps to learn how to interpret geologic cross sections by discussing and debating about how interpretations should be made.

- ☐ (2) Read the introduction to each cross section before beginning. They contain clues about what to look for in each cross section.

- ☐ (3) Develop a systematic strategy for analyzing each cross section. We suggest you begin with the steps below. Later you may develop your own strategy.

ANALYSIS OF CROSS SECTION

First, you must take the cross section apart and understand all the individual pieces. A well-analyzed cross section has notes written all over it.

STEP ONE   Locate and label all deformed sedimentary rocks by the law of original horizontality.

STEP TWO   Locate and label all unconformities. If you do not remember the unconformities, see page 138.

STEP THREE   Using the **law of cross-cutting relationships** examine all dikes and faults and number them in the order they occurred. Name each fault and label the force which causes it (tension, compression, shear).

STEP FOUR   Go to the bottom of the cross section and find the oldest rock unit by the **law of superposition** and label it the oldest. Number all the other rock units in order of occurrence by superposition and cross-cutting relationships.

STEP FIVE   Read the description of each rock unit and interpret the environmental conditions under which each forms. Write the conditions beside the rock on the cross section. Be sure to note rock units which are now in contact which could not possibly have formed under the same conditions; this means an unconformity is present. If you do not remember how a rock forms, go back to previous labs and look it up.

SYNTHESIS AND INTERPRETATION OF CROSS SECTION

Now that you have analyzed all the individual components of the cross sections, you can begin to put all the information together. Note that the page opposite each cross section is there for you to write out the steps in the geologic history of the cross section. The number of events is not important so long as the sequence is complete and in order.

STEP SIX   Go to the bottom of the diagram, find the oldest rock unit, and list it in the left-hand column as number 1 on the page opposite the cross section. In the right-hand column give the reason this unit comes first (in this case, it is superposition).

STEP SEVEN   Continue using superposition, original horizontality, and cross-cutting relationships to list the events in the history in their proper order.

OBSERVE: Every individual rock unit, each episode of faulting, and each igneous intrusion is, of course, an event and is numbered separately. But the following are also events and *these must also be included in the numbered sequence*: **periods of deformation** (horizontality violated), **unconformities** (period of uplift and erosion), **episodes of metamorphism** (burial and baking), and any other events which are distinct from the events above and below in the sequence.

OBSERVE: If you are not sure how to proceed, look at the example at the end of the Preliminary to Earth Deformation and Geologic History. Follow its pattern until you are able to do it on your own.

☐ (4) Ask your instructor to check your interpretations after each cross section. If you are making a mistake and keep repeating it, it will be more difficult to learn how to make correct interpretations.

INTRODUCTION TO CROSS SECTION NUMBER ONE (THE GRAND CANYON SECTION)

The geologic history exposed at the bottom of the Grand Canyon is long and intriguing, and the cross section below is typical of what you would see if you could take a trip down the Colorado River. It illustrates a number of classic geologic relationships which occur frequently all around the world. In particular, however, you will find three kinds of unconformities here. Since unconformities, among other things, are an integral and distinctive part of earth history, you should understand their meaning and learn how to recognize them on sight.

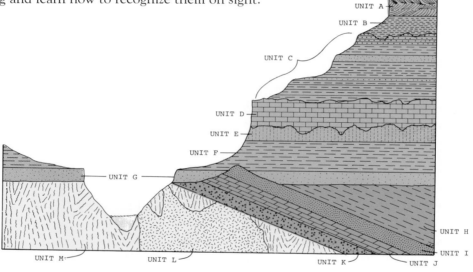

DESCRIPTION OF ROCK UNITS

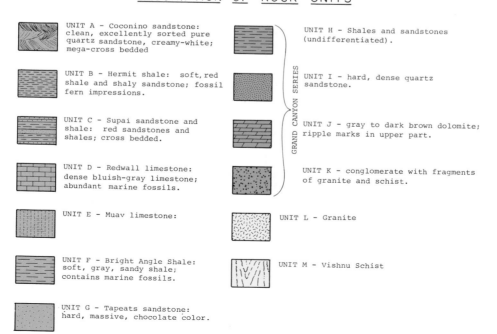

UNIT A - Coconino sandstone: clean, excellently sorted pure quartz sandstone, creamy-white; mega-cross bedded

UNIT B - Hermit shale: soft, red shale and shaly sandstone; fossil fern impressions.

UNIT C - Supai sandstone and shale: red sandstones and shales; cross bedded.

UNIT D - Redwall limestone: dense bluish-gray limestone; abundant marine fossils.

UNIT E - Muav limestone:

UNIT F - Bright Angle Shale: soft, gray, sandy shale; contains marine fossils.

UNIT G - Tapeats sandstone: hard, massive, chocolate color.

UNIT H - Shales and sandstones (undifferentiated).

UNIT I - hard, dense quartz sandstone.

UNIT J - gray to dark brown dolomite; ripple marks in upper part.

UNIT K - conglomerate with fragments of granite and schist.

UNIT L - Granite

UNIT M - Vishnu Schist

GRAND CANYON SERIES

GRAND CANYON SECTION

Event	Evidence, Law of Logic, and/or Interpretation

INTRODUCTION TO CROSS SECTION NUMBER TWO
(GREAT LAKES GEOLOGY)

The geologic history found in the Great Lakes Region of Minnesota, Northern Michigan, and Ontario is some of the oldest in the world. Many cycles of sediment accumulation, mountain building, regional metamorphism, and granite batholithic and other igneous intrusions have occurred here, each followed by erosion which removed the mountain to the sea. All this makes for interesting geology, much of which can be deciphered by recognizing unconformities and intrusive cross-cutting relationships (although it requires radioactive dating to sort out some of the more complex rock relationships).

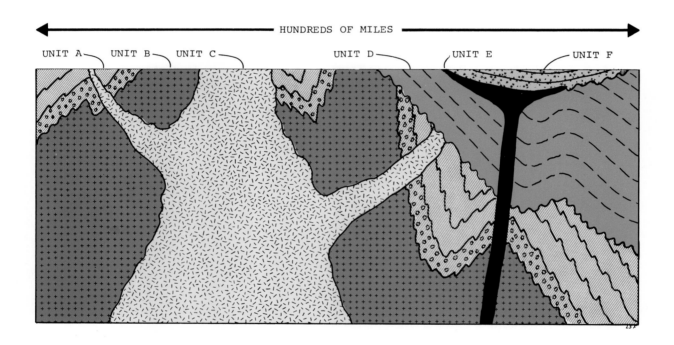

HUNDREDS OF MILES

UNIT A — UNIT B — UNIT C — UNIT D — UNIT E — UNIT F —

DESCRIPTION OF ROCK UNITS

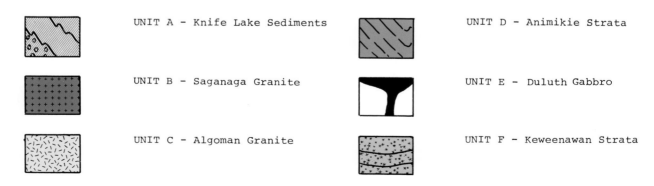

UNIT A - Knife Lake Sediments

UNIT B - Saganaga Granite

UNIT C - Algoman Granite

UNIT D - Animikie Strata

UNIT E - Duluth Gabbro

UNIT F - Keweenawan Strata

INTRODUCTION TO CROSS SECTION NUMBER THREE
(THE ABSAROKA MOUNTAIN SECTION)

This geologic cross section has been adapted (with some simplification) from a section of the geology found in the Absaroka Mountains of northwest Wyoming, near Yellowstone Park. It is, however, only a small part of the much more complex geology of this part of the world.

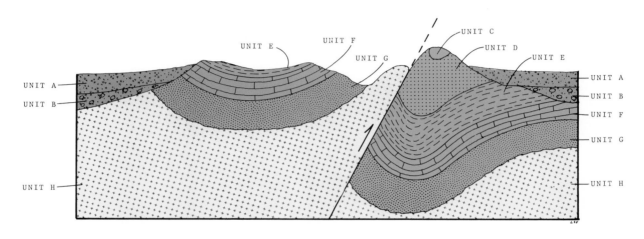

DESCRIPTION OF ROCK UNITS

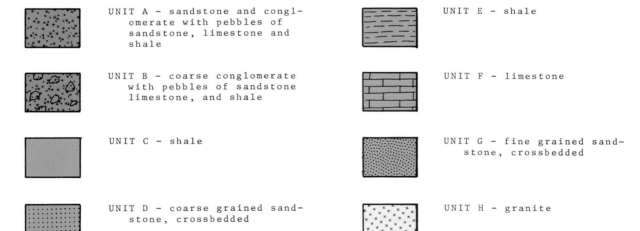

UNIT A - sandstone and conglomerate with pebbles of sandstone, limestone and shale

UNIT B - coarse conglomerate with pebbles of sandstone limestone, and shale

UNIT C - shale

UNIT D - coarse grained sandstone, crossbedded

UNIT E - shale

UNIT F - limestone

UNIT G - fine grained sandstone, crossbedded

UNIT H - granite

ABSAROKA MOUNTAIN SECTION

Event	Evidence, Law of Logic, and/or Interpretation

INTRODUCTION TO CROSS SECTION NUMBER FOUR

Although hypothetical, this cross section contains many of the structural elements found in the mountain systems of the western United States. It represents one of the typical patterns of structural deformation; that is, compression has preceded tension. The solution to this geologic history lies in unraveling the sequence of compression (folding and faulting) and tension (faulting) via the law of cross-cutting relationships, among other evidence.

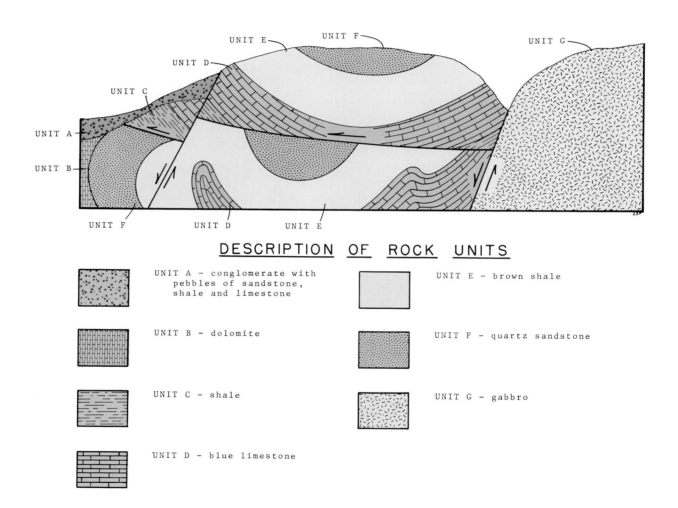

DESCRIPTION OF ROCK UNITS

UNIT A – conglomerate with pebbles of sandstone, shale and limestone

UNIT B – dolomite

UNIT C – shale

UNIT D – blue limestone

UNIT E – brown shale

UNIT F – quartz sandstone

UNIT G – gabbro

CROSS-SECTION NUMBER FOUR

Event	Evidence, Law of Logic, and/or Interpretation

INTRODUCTION TO CROSS SECTION NUMBER FIVE
(TRIASSIC BASINS OF EASTERN NORTH AMERICA)

This cross section is adapted from the kinds of "fault basins" which developed during the Triassic Period in the Piedmont of eastern North America. The rock types and their relationships are typical responses to the processes occurring during a rifting event and the opening of a new ocean basin. The faults causing the basin result from tension but do not develop all at once. Rather they develop as numerous small movements which progressively deepen the basin, while between movements sediment is washed into the depression. Through this alternate faulting and sediment filling, the basin develops a thick wedge of sediment, thickest near the faults.

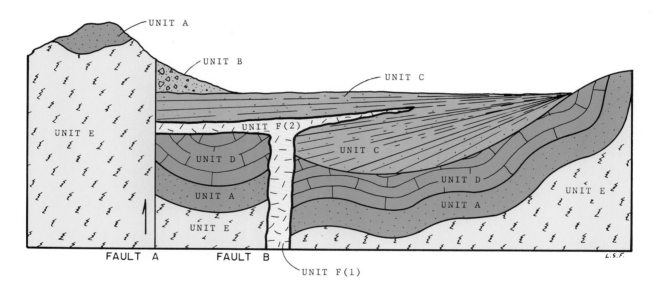

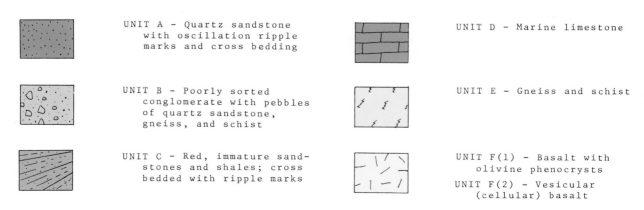

DESCRIPTION OF ROCK UNITS

UNIT A - Quartz sandstone with oscillation ripple marks and cross bedding

UNIT B - Poorly sorted conglomerate with pebbles of quartz sandstone, gneiss, and schist

UNIT C - Red, immature sandstones and shales; cross bedded with ripple marks

UNIT D - Marine limestone

UNIT E - Gneiss and schist

UNIT F(1) - Basalt with olivine phenocrysts

UNIT F(2) - Vesicular (cellular) basalt

TRIASSIC BASINS OF EASTERN NORTH AMERICA

Event	Evidence, Law of Logic, and/or Interpretation

INTRODUCTION TO CROSS SECTION NUMBER SIX
(MEDICINE BOW MOUNTAINS)

In the Rocky Mountains of western North America, beginning in the late Cretaceous and continuing through much of the early Cenozoic, a series of intermountain depositional basins began to develop, surrounded by north- or northwest- trending mountain ranges. The patterns of developing mountain ranges and associated sediments deposited in the intermountain basins are typical of the region and illustrate the strong control **tectonics** (earth movement) has on sediment associations. The state of Wyoming best typifies this type of geology, and the cross section here has been adapted (with some simplification for clarity) from the Medicine Bow Mountains of southeastern Wyoming.

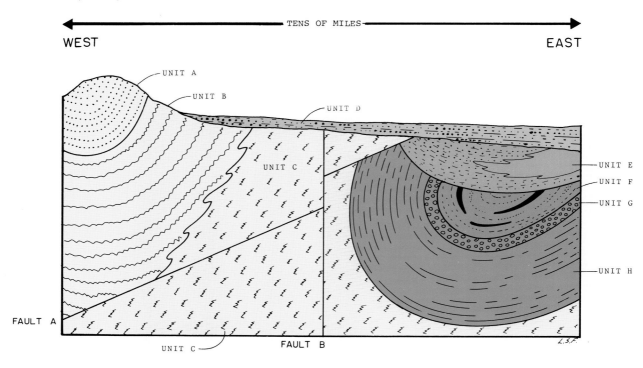

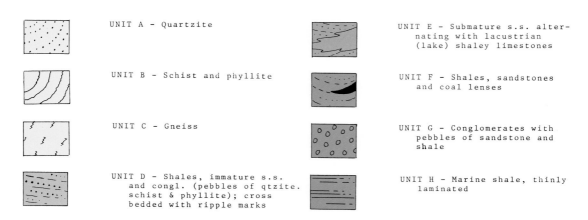

DESCRIPTION OF ROCK UNITS

UNIT A - Quartzite

UNIT B - Schist and phyllite

UNIT C - Gneiss

UNIT D - Shales, immature s.s. and congl. (pebbles of qtzite. schist & phyllite); cross bedded with ripple marks

UNIT E - Submature s.s. alternating with lacustrian (lake) shaley limestones

UNIT F - Shales, sandstones and coal lenses

UNIT G - Conglomerates with pebbles of sandstone and shale

UNIT H - Marine shale, thinly laminated

MEDICINE BOW MOUNTAINS

Event	Evidence, Law of Logic, and/or Interpretation

INTRODUCTION TO CROSS SECTION NUMBER SEVEN

The situation represented here is more complex than the previous six, and although not directly adapted from any particular locality, it illustrates the relationships which developed in the formation of folded and thrust-faulted mountain ranges of western North America. In many complex geologic circumstances, rock relationships are hard to decipher without having good information on their relative ages; this is the case here. Since you do not have this information, several correct interpretations are possible. Just using simple logic, however, will sort out an acceptable solution. To make the interpretations easier, consider the rocks found in each sequence (A, B, C, and D) as a unit, formed at the same relative time.

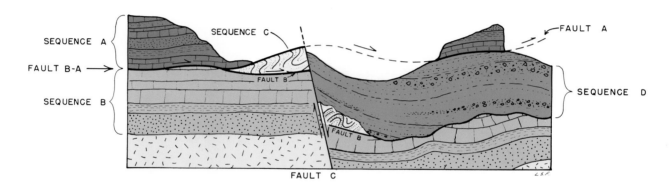

DESCRIPTION OF ROCK UNITS

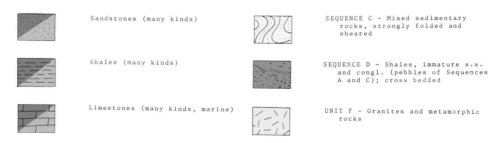

Sandstones (many kinds)

Shales (many kinds)

Limestones (many kinds, marine)

SEQUENCE C - Mixed sedimentary rocks, strongly folded and sheared

SEQUENCE D - Shales, immature s.s. and congl. (pebbles of Sequences A and C); cross bedded

UNIT F - Granites and metamorphic rocks

CROSS SECTION NUMBER SEVEN

Event	Evidence, Law of Logic, and/or Interpretation

3-D BLOCK DIAGRAMS

On the following pages are the 3–D block diagrams referred to on pages 145 and 151.

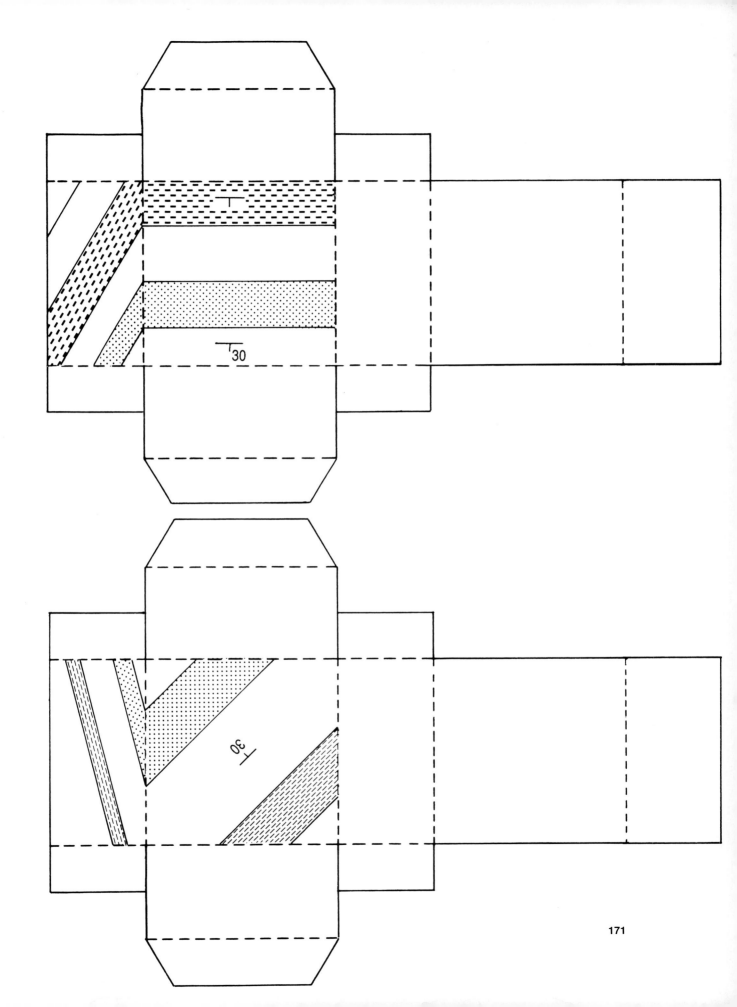

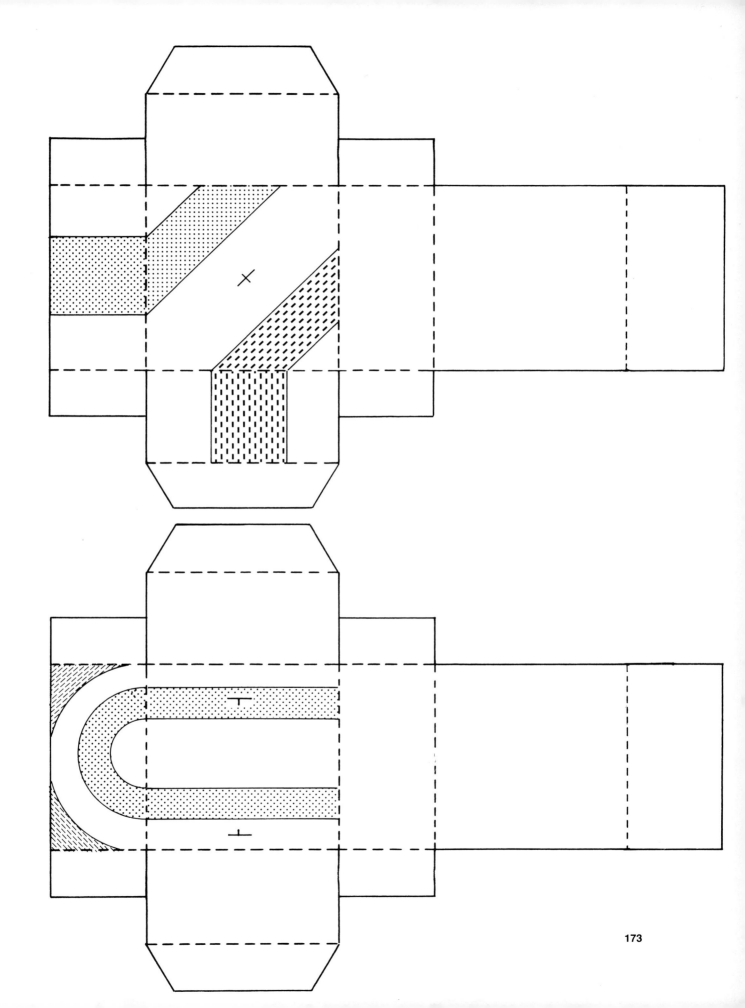

173

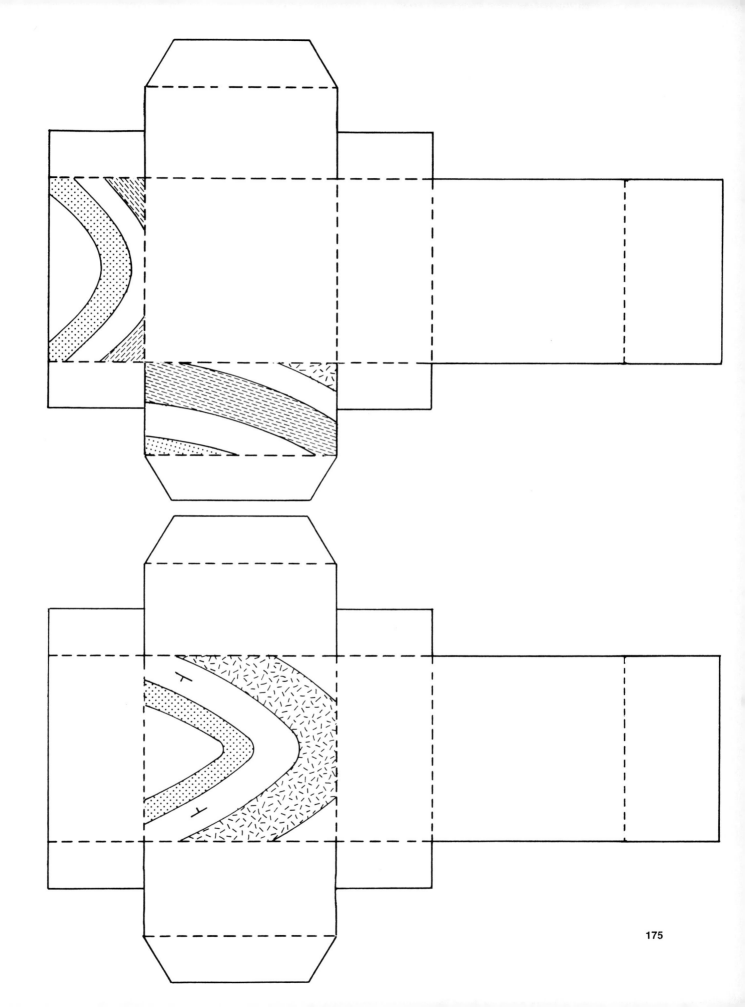

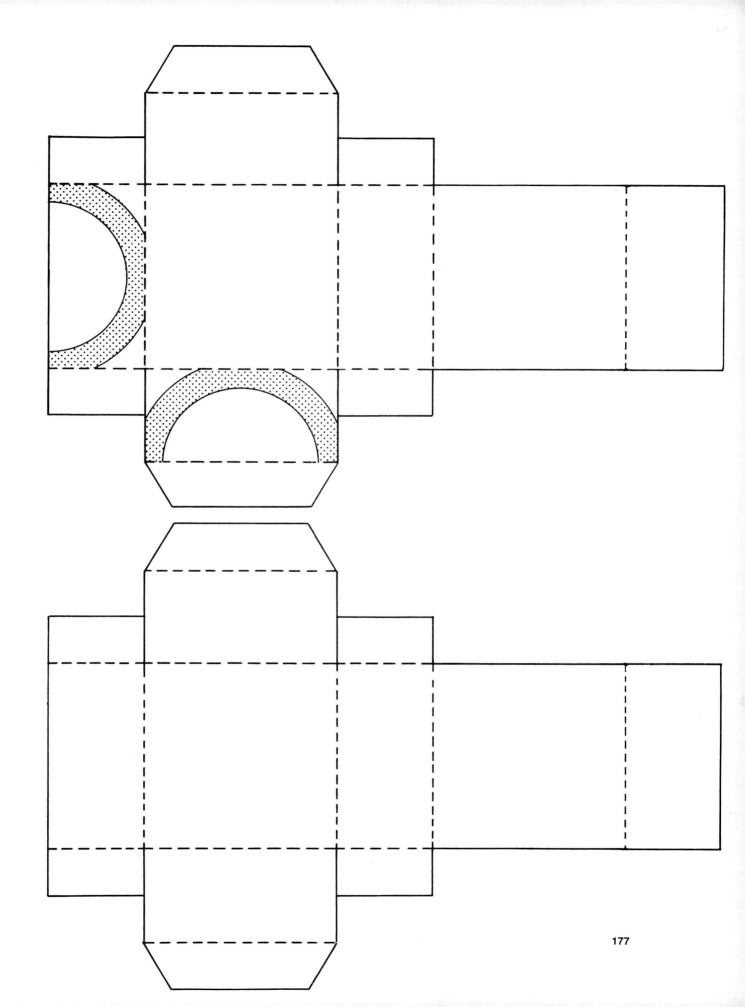

177

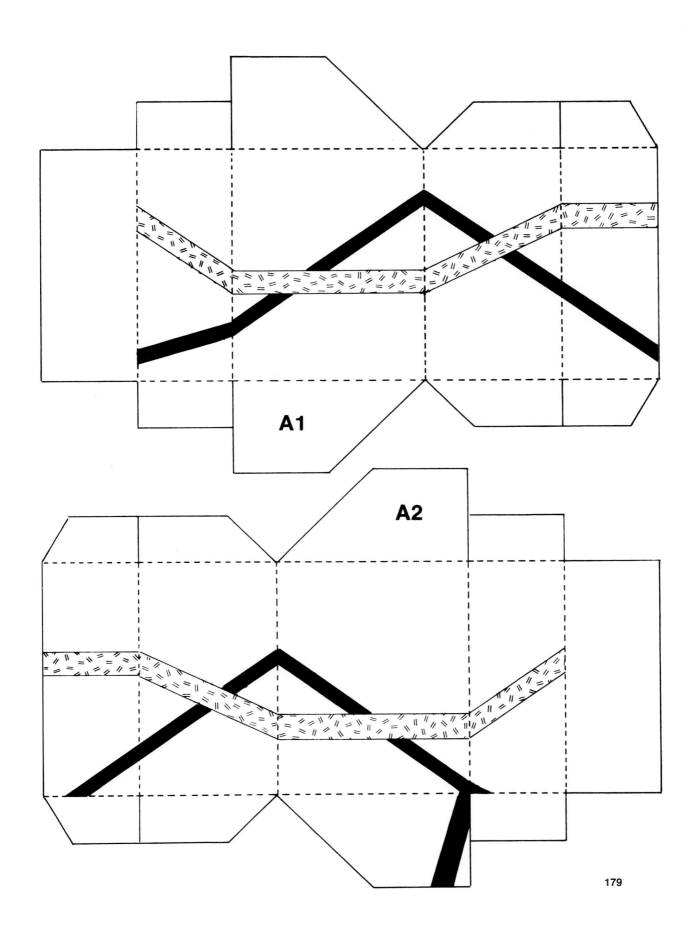

A1

A2

179

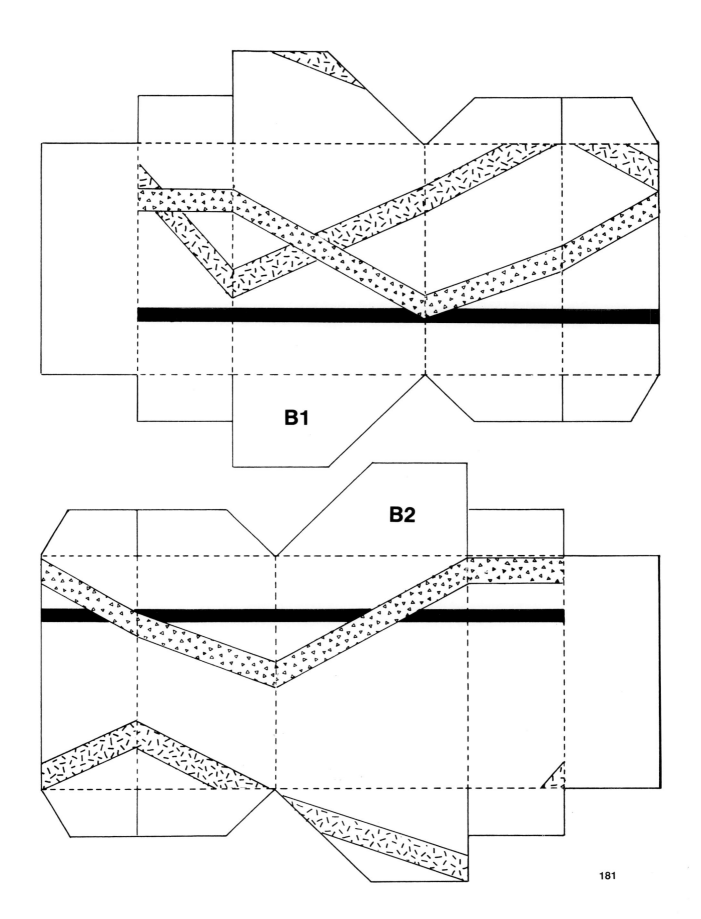

B1

B2

181

Geologic Maps

PURPOSE

Studying geologic maps after cross sections is the opposite to the order in which geologists do these studies. When a geologist enters an area for the first time, the first work that must be done is to draw a geologic map. Geologic cross sections are interpretations then drawn from the information on the geologic map. We had you interpret geologic cross sections first (lab on *Earth Deformation and Geologic History*) because it is much easier to visualize and learn structural geology from cross sections. But now it is time to interpret geologic maps and learn how to draw cross sections from them. Interpreting the geologic maps will require that you take all the knowledge and skills you have acquired so far and apply them to the interpretation of rocks and structures as they are exposed at the earth's surface, and then project that knowledge to an interpretation of what is happening inside the earth where we cannot go.

This laboratory has just one part.

INTRODUCTION

Most of the rules and methods used to interpret geologic history from cross sections apply to geologic maps, but three special features are seen on or apply only to geologic maps. These you must learn before venturing to interpret the geologic maps.

HOW OUTCROP WIDTH VARIES ON A GEOLOGIC MAP   Outcrop width refers to the area taken up by a rock unit when seen from the air, or on a map. From the air a flat lying, or horizontal, bed of rock seems to blanket everything and has great outcrop width, even if the bed is in fact very thin. The same thin bed of rock oriented vertically would appear as a line when seen from the air. In between horizontal and vertical, the outcrop width can vary from narrow to very broad.

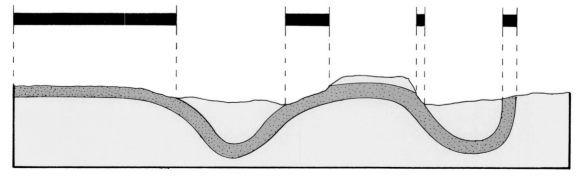

One caution about interpreting outcrop width is that large bodies of rocklike igneous batholiths may *appear* to have great outcrop width, like horizontal sedimentary beds, when in fact they are massive bodies of rock which would not vary in thickness no matter how they were oriented.

OUTCROP PATTERNS ON GEOLOGIC MAPS   Horizontal sedimentary rocks and large exposures of igneous and metamorphic rock rarely show any distinctive outcrop pattern. Either they just cover a large area, or they have an irregular outcrop pattern without noticeable patterns. Folded sedimentary rocks (or any other layered rock) are quite different in that they always have a distinctive outcrop pattern that can be instantly recognized. The two most important of these are anticlines (domes) and synclines (basins)[1] (see illustrations below).

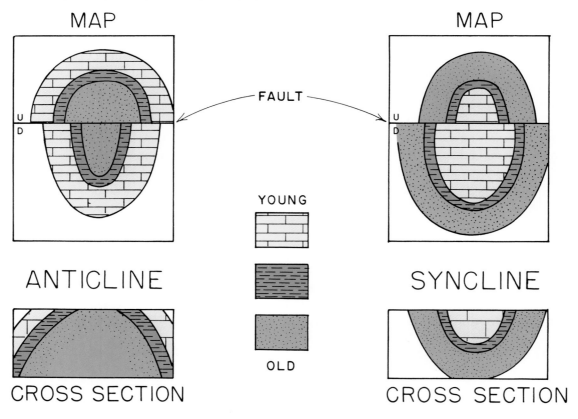

When exposed at the surface and eroded, both anticlines and synclines have a concentric banded pattern. Although they look alike at first, they differ in three fundamental ways.

1. **Anticlines** when eroded expose the oldest beds in the center, and the beds become progressively younger toward the outside. **Synclines,** on the other hand, have the oldest beds exposed on the outside, and the beds become progressively younger toward the center.

2. The beds in an anticline dip *away from the center*, and therefore the anticline gets wider with depth. This means that as an anticline is eroded its area of exposure at the surface becomes larger.
    The beds in a syncline dip *toward the center*, and therefore the syncline gets narrower with depth. This means that as a syncline is eroded, its area of exposure at the surface becomes narrower.

[1] Anticlines which are nearly circular in outline are called **domes.** Synclines which are nearly circular in outline are called **basins.**

3. If an anticline is faulted, the outcrop pattern on the side that has moved up (upthrown side) increases in width relative to the side which has not moved. In a syncline which has been faulted, the outcrop pattern on the upthrown side decreases in width relative to the side that has not moved (see illustration page 184).

THE RULES OF "V"   If you were to trace, from the air, the outcrop pattern of a perfectly straight (that is, unfolded) but dipping bed of rock across flat land, it would appear as a straight line of outcrop—until it crosses a valley. As the bed crosses the valley, its outcrop pattern will become "V-shaped." This is the basis for the Rules of "V."

FIRST RULE   When any dipping plane (bed, dike, fault, etc.) crosses a valley, the outcrop pattern will form a "V" shape. In most cases *the "V" points in the direction the bed is dipping*.

The Rules of "V" are very useful because they indicate the direction a bed, dike, fault plain, and so on is dipping just by looking at the outcrop pattern on a geologic map without going out to see the rock. There are three exceptions to the rule.

SECOND RULE   A vertical plane *never* shows a "V."

THIRD RULE   Horizontal rock units *do* show a "V" even though they are not dipping. These cases can usually be recognized because the horizontal beds will blanket large areas of the map. Also, each valley forms its own "V" and several different valleys may "V" in different directions, implying mutually contradictory dips. This is a clue to a flat-lying bed.

FOURTH RULE   There is one case where the "V" points in the opposite direction of the dip. This happens only when the angle of slope of the valley floor is greater than the angle of dip of the bed.

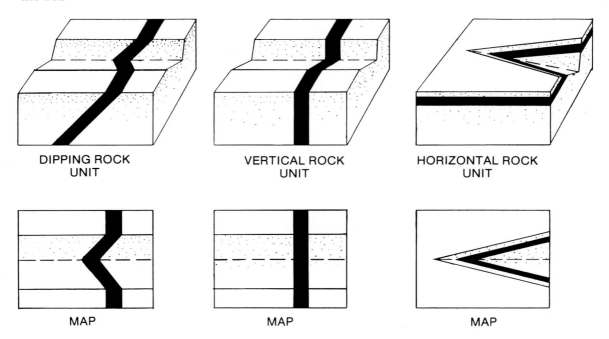

DIPPING ROCK
UNIT

VERTICAL ROCK
UNIT

HORIZONTAL ROCK
UNIT

MAP   MAP   MAP

CONSTRUCTING GEOLOGIC CROSS SECTIONS

In this section you will learn to draw geologic cross sections from geologic maps. There are two ways to approach this, one haphazard and disorganized, the other deliberate and systematic. At first it takes more time and effort to learn to draw cross sections deliberately and systematically, but:

> **OBSERVE:** We can virtually guarantee you that if you try to draw geologic cross sections from geologic maps in a haphazard and disorganized way, you will never get them right. One of the most important things you can learn here is that the solution of scientific problems must be done in as deliberate and systematic manner as possible.

The object of these exercises is not to just get them done, it is to cultivate the habit of developing a strategy for solving complex problems in a deliberate and systematic manner. A deliberate and systematic analysis means that you first take the map apart and understand all the individual pieces. And then you synthesize all the pieces into a whole interpretation. A well-analyzed map has notes written all over it. The example below shows one way to analyze a geologic map and draw a cross section. We have done the analysis in a series of steps which you can use as you learn to do your analyses. In time you will develop your own strategy of analysis.

AN EXAMPLE OF AN ANALYSIS

This analysis is for the geologic map and cross section below:

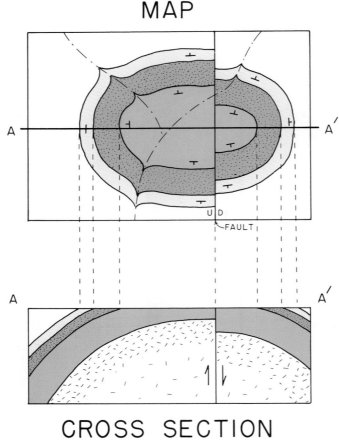

MAP

CROSS SECTION

— · — · — STREAMS

STEP ONE   Draw strike and dip symbols on every rock unit and every fault plane which does not already have them by using the Rules of "V."

STEP TWO   Label every structure you can identify. In the example below,

1. **Anticline.** The concentric outcrop pattern tells us a fold is present. The strike and dip symbols tell us the beds dip away from the center so the fold is an anticline (dome).

2. **Vertical fault.** Because the beds of the anticline are offset, a fault is present. A stream valley crosses the fault but no "V" is formed by the fault plane; thus, the *fault plane* is vertical. The fault could be a lateral fault if all the beds of the fault were offset in the same direction; the fact that they are not tells us the fault itself is vertical and not lateral.

Sometimes you will have to use rock type to make identifications. For example, a sill runs parallel to sedimentary bedding but is recognized because it is composed of igneous rock, not sedimentary.

STEP THREE   Project all contacts, faults, and any other structures from the cross section line A-A' down onto the cross section. *A hint:* For each contact you project down to the cross section, draw a little arrow pointing the direction that contact dips *into* the cross section.

OBSERVE: Avoid these two common mistakes. Even experienced people make these mistakes if they are not paying attention. Not paying attention and making one or both mistakes will guarantee that your cross section will be wrong.

First mistake: Projecting a contact from the map to the cross section from some line other than the cross section line, such as the bottom line of the map. *Solutions*: Double-check every contact you project. Also, cover the map above the cross-section line with a piece of paper so that you are forced to focus on the line of cross section.

Second mistake: Confusing the angle a rock unit takes across the map with the dip of the rock unit. For example, Geologic Map Two on page 190 has a dike running across it diagonally (Unit D). Beginners often want to make such a rock unit dip in the cross section too, but notice that on the map no "V"s are formed where streams cross the dike. The dike is vertical and must be drawn straight down on the cross section. *Solution:* Deliberately double-check the dip of every bed before you project it down to the cross section.

STEP FOUR   Make your interpretation by sketching the structures below the surface on the cross section.

OBSERVE: Drawing realistic cross sections takes practice and guidance. Draw the best cross section you can, and then ask your instructor to make suggestions on how it can be made better.

DRAWING YOUR OWN GEOLOGIC CROSS SECTIONS

☐   (1)   Work together in groups of two or three. It helps to learn how to draw geologic cross sections by discussing and debating about how interpretations should be made.

☐   (2)   Analyze the structures on the map. Use the steps in the example above as a beginning.

☐ (3) Ask your instructor to check your cross section after it is completed.

☐ (4) Using the geologic map and your cross section list the history in the region of the map from first to last. Follow the procedures used in the laboratory on *Earth Deformation and Geologic History* (page 155).

INTRODUCTION—GEOLOGIC MAP ONE

Although hypothetical, the area below is representative of the geology found in major folded mountain systems. It illustrates the classic pattern of folded rock outcrop exposed at the earth's surface and the topography which results.

MAP

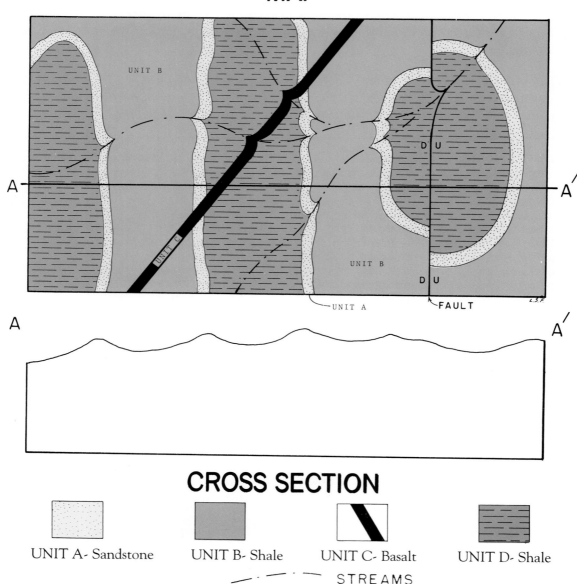

CROSS SECTION

UNIT A- Sandstone   UNIT B- Shale   UNIT C- Basalt   UNIT D- Shale

STREAMS

Event	Evidence, Law of Logic, and/or Interpretation

INTRODUCTION—GEOLOGIC MAP TWO

The area below has had a long and complex geologic history, which has included tension, compression, intrusion, metamorphism, erosion, and so on. It represents a pattern classically developed in certain parts of the world.

MAP

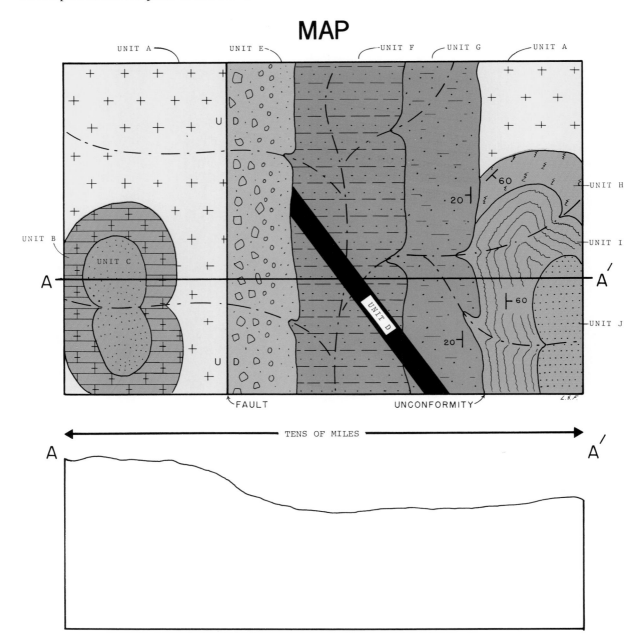

CROSS SECTION

— · — · — · STREAMS

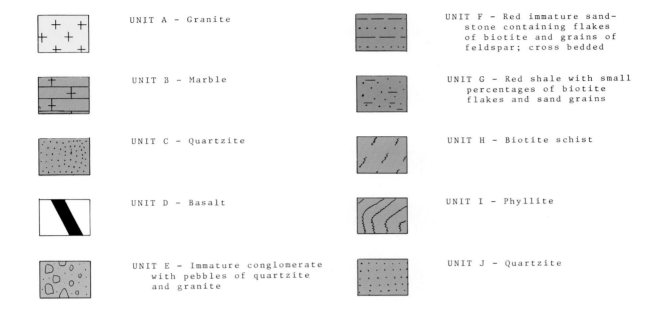

UNIT A - Granite

UNIT B - Marble

UNIT C - Quartzite

UNIT D - Basalt

UNIT E - Immature conglomerate
with pebbles of quartzite
and granite

UNIT F - Red immature sandstone containing flakes
of biotite and grains of
feldspar; cross bedded

UNIT G - Red shale with small
percentages of biotite
flakes and sand grains

UNIT H - Biotite schist

UNIT I - Phyllite

UNIT J - Quartzite

Event	Evidence, Law of Logic, and/or Interpretation

INTRODUCTION—GEOLOGIC MAP THREE

The region in the map below is typical of the folded and thrust-faulted mountain systems which have developed in many parts of the world. It is taken from the geology found in the Shenandoah Valley of Virginia, and although it may have been simplified in places it is accurate in all essential features.

Two sections have been provided —A-A' and B-B'—to illustrate some of the variations possible.

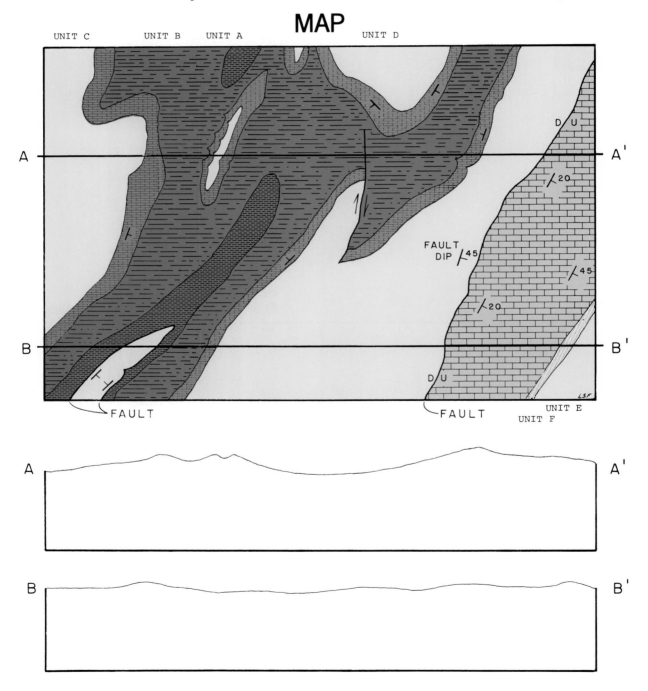

DESCRIPTION OF ROCK UNITS

 UNIT A - Martinsburg: shales and silty-shales

 UNIT D - Beekmantown: cherty dolomite

 UNIT B - Edinburg: shales and black limestones

 UNIT E - Chepultepec: blue limestones

 UNIT C - Lincolnshire— New Market light gray limestone and dark gray cherty limestone

 UNIT F - Conococheaque: thin bedded blue limestone and gray dolomite

Event	Evidence, Law of Logic, and/or Interpretation

INTRODUCTION—GEOLOGIC MAP FOUR

Although hypothetical, the geologic relationships illustrated below are like those found in regions of the world which have undergone periods of extensive mountain building. The mountains are now gone, having been eroded away, but their roots can now be seen at the earth's surface. The rocks typically exposed in these ancient mountain regions show regional metamorphism with several episodes of igneous intrusion.

MAP

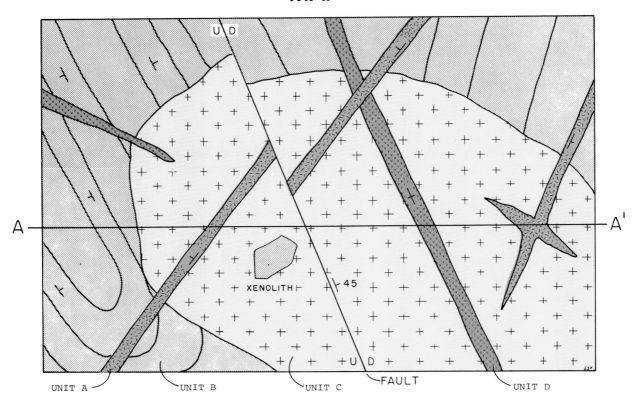

CROSS SECTION

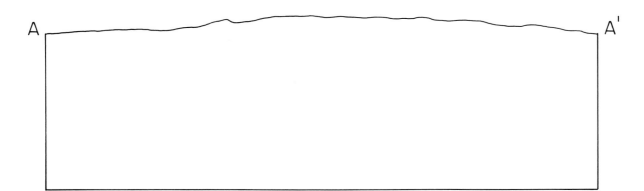

Event	Evidence, Law of Logic, and/or Interpretation

Preliminary to a Plate Tectonic Rock Cycle

PURPOSE

In each of the previous labs we have been systematically developing the idea that every rock forms under a specific and unique set of circumstances. Find the rock and you know what those circumstances were at that spot at the time that particular rock formed. Change the circumstances and a different rock forms. Much of the work of a professional geologist involves trying to understand these relationships—and in full dress it is exceedingly technical and complex. Yet the basic relationships are now well enough known they can be understood by anyone.

Also in each of the previous labs we have been introducing, piece by piece, ideas about how the earth is structured and functions. In this lab we draw all these ideas together into one synthetic model. You will probably find many of the ideas in the model familiar, and now you will see how all the processes and pieces lock together to form a very dynamic theory of the earth. By the time you enter the next laboratory you should be able to:

1. Explain what is meant by **plate tectonic theory.**

2. Be able to sketch and explain the **rock cycle.**

3. Distinguish between **island arc–type** and **cordilleran-type subduction zones.**

4. Label on a cross section of the earth the following features: **forearc, backarc, remnant ocean basin, paired metamorphic belt (blueschist** and **regional), foreland, hinterland,** and **suture zone.**

INTRODUCTION

No rock is accidental. No idea in geology is more profound than this; it runs from the center to the whole of geology and influences every subdiscipline of the field. Genuine understanding of the science of geology begins with one's ability to understand and explain why no rock is accidental.

THE ROCK CYCLE The rock cycle is the starting point for understanding a theory of the earth for it summarizes how one rock can change into another (see figure on page 198). In each laboratory so far, rocks have been explained in terms of the processes by which they form. We see that an igneous rock exposed at the earth's surface is not stable but will weather into sedimentary minerals which are stable at the earth's surface. These weathering products evolve as they are transported downstream, but eventually in the depositional basin, they are buried to a depth where heat and pressure metamorphose them. As the metamorphism increases, the rock changes from slate, to phyllite, to schist, to gneiss, and finally to magma—and we are back where we started with an igneous rock. From studies of the oldest rocks, dated nearly 4 billion years old, we know that these processes have been part of the earth from the beginning.

THE ROCK CYCLE [1]

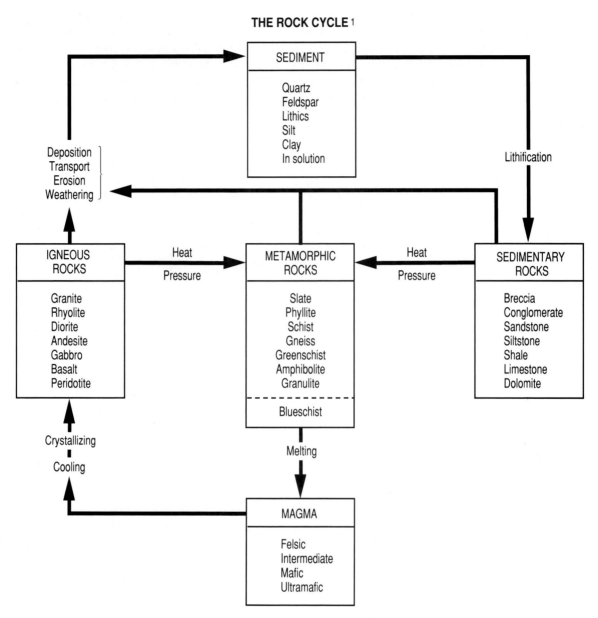

PLATE TECTONICS   **Tectonics** is concerned with deformation in the earth and the forces which produce deformation.

Plate tectonics is the theory that the earth's lithosphere (outer rigid shell) is composed of several dozen "plates," or pieces, which float on a ductile mantle, like slabs of ice on a pond. In plate tectonic theory, earth history, at its simplest, is one of plates rifting into pieces diverging apart and new ocean basins being born, followed by motion reversal, convergence back together, plate collision, and mountain building.

Plate tectonics is one of the great unifying theories in geology. Virtually every part of the earth's crust, and every kind of rock, and every kind of geology, can be related to the plate tectonic conditions which existed at the time they formed. Everything in geology makes more sense in terms of plate tectonic theory.

[1] Thanks to Howard Campbell for the original idea for this style rock cycle.

The history and evidence for plate tectonic theory continues to grow and evolve. The scientific tests it has survived, and which support its validity, are rich and powerful—studies utilizing gravity, paleomagnetics, earthquake activity, volcanic activity, fossil distribution, and so on. Our purpose here is not to review the evidence and support for plate tectonic theory. Your lecture instructor and text book will explore that. We will assume that plate tectonic theory is valid and use it to understand those processes which explain how specific structures and rocks form within the earth.

EARTH STRUCTURE

In cross section the earth is divided into three layers: a core, mantle, and lithosphere. The core is the molten center of the earth and does not interest us here. The mantle is made of hot, ductile ultramafic rock (peridotite and related rocks) which flows very slowly (rates of centimeters per year) in convection cells circulating the mantle material over and over. The lithosphere is a relatively thin layer of lightweight, brittle rock floating on the underlying ductile mantle(see illustration).

The lowermost layer of the lithosphere is made of ultramafic rock (for example, peridotite) which is, in fact, the cool, brittle, upper-most portion of the mantle. Above the ultramafic rock is the crust composed of one of two lower density rocks floating isostatically,[1] either basalt/gabbro or granite.

The lithosphere is subdivided in two ways. One way of subdividing the lithosphere is into ocean basins and continents. Ocean basins are made of mafic igneous rocks (basalt and gabbro) which, because they are dense, float isostatically on the mantle at a depth of nearly 6 km below sea level. Continents are made of granites or granite-like rocks, which, because they are of lower density, float isostatically near sea level.

The second way of subdividing the lithosphere is into **plates**. The lithosphere today is made of about a dozen major plates and many minor ones. A plate can be made of oceanic crust alone, but a plate with a continent virtually always includes oceanic crust attached to the continent.

Most geologic activity occurs at the boundaries between plates. There are three kinds of boundaries: (1) **divergent boundaries,** where plates are moving apart and new crust is being created; (2) **convergent boundaries,** where plates are moving together and crust is being destroyed; and (3) **transform boundaries,** where plates slide past one another. Very interesting geology occurs along transform boundaries, as the faulting along the San Andreas fault system in California attests to, but our model will not discuss transform boundaries.

The following model begins with a hypothetical, geologically quiet, earth. The model is divided into ten stages, but the stages are arbitrary and do not exist naturally. The earth is an ongoing series of processes, so it is much more important to understand the processes, how they are related, and how one process leads naturally to the next process than to try to memorize arbitrary stages.

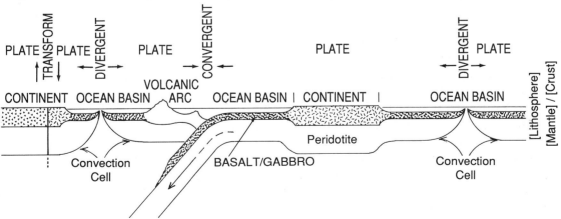

[1] Something in isostatic balance is floating in equilibrium , like a boat resting in water. Nothing is artificially holding it up or pushing it down.

THE MODEL: A PLATE TECTONIC ROCK CYCLE

The most important message of the rock cycle is that beginning with an igneous rock of mafic composition (e.g., basalt), all the other rocks now on the earth can be generated. The most important message of the plate tectonic rock cycle is that each and every rock forms under a specific set of tectonic conditions.

☐   (1)   Get the chart titled "A Plate Tectonic Rock Cycle" from the pocket at the back of the manual and lay it open on the table. It contains ten geologic cross sections corresponding to the model discussed below.

☐   (2)   Read the Model: A Plate Tectonic Rock Cycle which begins below and carefully study the illustrations on the chart "A Plate Tectonic Rock Cycle" as you read. The model is as much visual as verbal so take the time to study the cross sections until you begin to see small details in them.

STAGE 1—A STABLE CONTINENTAL CRATON

Imagine a very simple situation—a small tectonically stable continental craton bordered by ocean basins all around. The continent is eroded down to sea level everywhere (a **peneplain**); it is dead flat from edge to edge and corner to corner and there is no tectonic activity at all. On the surface is a blanket of supermature quartz sandstone, the result of millions of years of weathering and sorting. Limestone may be deposited also, but most shales (clays) have been blown or washed off the continent into the ocean basins over the millions of years of stability. The continent is in perfect isostatic equilibrium; by itself it will not rise or sink. Nothing exciting or interesting is happening; no earthquakes or volcanic activity. The situation is unrelenting boredom.

STAGE 2—HOT SPOT AND RIFTING

Into the peaceful stable continent comes a disturbance. From deep in the mantle a hot spot, a plume of hot mafic magma rises toward the surface and ponds at the base of the continent. The hot spot heats the base of the continental crust, causing it to thin and stretch like pulled taffy (or silly putty) and then swell into a large dome maybe 1,000 km in diameter. The expanding dome stretches across the top, cracks into a series of normal faults, and collapses in the center into an **axial rift** (also called an **axial graben**). The axial rift is tens of kilometers across, and the elevation from the floor of the rift to the mountain crests on either side (**horsts**) is as much as 4–5 km.

Rifting is splitting the original continent into two pieces, west and east, but they are still connected. Mafic volcanoes are common and appear as cinder cones and flood basalts in the rift. Sometimes a very intense hot spot will melt the lower continental (granitic) crust, and large felsic volcanoes will form.

The horsts weather and erode, and large volumes of coarse arkosic breccias and conglomerates are deposited on the sides of the axial rift. In the center of the rift, lakes are common and fill with deep-water, black, fine-grained sediments. In time, as the continent stretches apart even more, the axial graben drops below sea level, and marine water invades the rift, forming a narrow sea, like the modern Red Sea or Gulf of Aden.

STAGE 3 — CREATION OF NEW OCEANIC CRUST AND AN EARLY DIVERGENT MARGIN

Now a great surge of volcanic activity begins along one side of the axial rift, in this case the east side. At first the magma is injected as a large number of basaltic dikes into the granitic continental crust, so many dikes in fact that it is finally hard to decide what the original rock was. This mixture of granite with basalt dikes is called **transition crust.**

As the volcanic activity continues, the east and west pieces of the original continent drift apart and the gap between them fills with mafic igneous rock. Surge after surge of magma rises from a convection cell in the mantle into the continuously spreading gap as the continents move farther and farther apart. Because all this new igneous rock is mafic in composition (basalt near the surface and gabbro at depth) and high in density, it floats isostatically below sea level, creating new oceanic crust and a new ocean basin. Within a few million years, the two continents can be separated by thousands of kilometers.

The final result is that beginning with only one tectonic plate in stage 1, rifting has created a new divergent plate boundary and two plates, one on the west containing Westcontinent and one on the east containing Eastcontinent.

STAGE 4—FULL DIVERGENT MARGIN

The Eastcontinent has now drifted off the eastern side of the cross section, and only the Westcontinent and new ocean basin remain to be studied. Heat rising to the surface from the convection cells remains concentrated at the rifting site in the new ocean center, so as the ocean basin widens the newly formed continental margins move away from the heat source and cool. Cool crust is denser than warm crust, and the divergent margin sinks below sea level, rapidly at first, but ever more slowly until it stabilizes after 100 million years.

Meanwhile a great wedge of sediment eroding from the continent accumulates on the divergent continental margin (also called a **passive continental margin** because it is geologically passive or an **Atlantic-type** margin because it is like the margins at the Atlantic ocean today). It is mostly shallow-water marine deposits because subsidence and deposition go on at about the same rate. When next to a stable craton, the wedge of sedimentary rocks is dominated by mature sandstone, limestone, and dolomites, but if the continent is tectonically active, many kinds of less mature sedimentary rocks are possible.

STAGE 5—CREATING A CONVERGENT BOUNDARY: VOLCANIC ISLAND ARC MOUNTAIN BUILDING

Divergence, and the creation of new oceanic crust, can go on for tens or hundreds of millions of years. At some point, however, divergence stops and the two continents begin to move back toward each other. This is convergence, and a new plate boundary must be created for it. Convergence begins when oceanic crust **decouples**, that is, breaks at some place and begins to descend into the mantle along a **subduction zone.**

It is always oceanic crust which decouples and descends into a subduction zone; continental crust is too light to subduct. Subduction zones can form anywhere in the ocean basin. In the stage 5 cross section subduction is dipping east, but it could have been west, or any direction. There are just two kinds of locations for subduction zones, however: one within an ocean basin (island arc type), the other along the edge of a continent (cordilleran type). Both kinds of subduction cause volcanic mountain building, and they are extremely important. Things are heating up now compared to the boredom of stage 1. The island arc type is described next; the cordilleran type later.

Subduction Orogenies

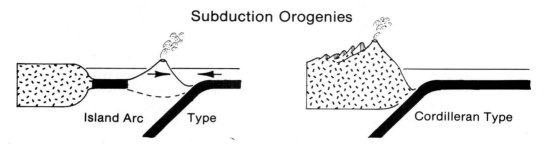

Island Arc   Type   Cordilleran Type

EARLY VOLCANIC ISLAND ARC MOUNTAIN BUILDING At a subduction zone oceanic crust dives into the mantle. When oceanic crust subducts, it sets in motion a chain of processes which creates several new structural features and generates several new kinds of rocks.

Structural Features At the site of subduction, part of the oceanic crust is dragged down into a **trench** 1–2 km below the ocean floor. The subducting oceanic crust begins its descent cold but heats up as it slides into the mantle. By 120 km deep the rock is hot enough to begin melting to form magma. The magma rises toward the surface, forms batholiths, breaks onto the ocean floor, and builds a volcano which eventually rises high enough to form an island. The location of the volcano is called the **volcanic front** (in three dimensions it is a string of volcanoes all rising above the subduction zone). The area on the trench side of the volcanic front is the **forearc**, and the area on the back side of the volcanic front is the **backarc**. A new convergent boundary has been created along the zone of subduction.

Fractional Melting and the Creation of New Igneous Rocks Subducting oceanic crust selectively melts and fractionates. In fractional melting an igneous rock of one composition is divided into two fractions each of a different composition. The first fraction is a melt whose composition is more felsic than the original rock. That is, its composition is closer to the bottom of Bowen's Reaction Series than the original rock (see pages 24–26; *Preliminary to Igneous Rocks*). The second fraction is a residue with a composition more mafic than the original rock. That is, its composition is higher in Bowen's Reaction Series than the original rock. For example, in a subducting oceanic crust of mafic composition, the fractionation is:

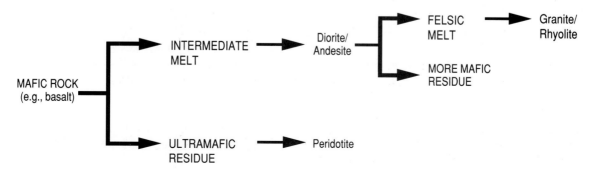

If time and conditions allow, the fractionation process can continue and the intermediate magma can fractionate into felsic magma, leaving behind a residue more mafic than the felsic melt. Thus, beginning with one (mafic) igneous rock, many new igneous rocks can be generated, including ultramafic, intermediate, and felsic. An important way to think about this is that felsic continental crust is created from the fractional melting of mafic oceanic crust.

In our subduction zone, the ultramafic residue, being very dense, stays in the mantle, while the hot, less dense, melt rises to the surface where it forms first intermediate and later felsic batholithic magma chambers, From the chamber, the magma reaches the surface as lava and forms volcanoes, which are dominated by andesite, although it can evolve from mafic, to intermediate, to felsic as the magma fractionates. Hydrothermal metamorphism also occurs when hot lava spills out onto the ocean floor and reacts with cold seawater.

Remnant Oceans   Now, step back and look at the whole of cross section 5. Notice that the ocean basin to the west of the convergent plate boundary is trapped between a divergent continental margin (eastern edge of Westcontinent) and the subduction zone. Clearly, if subduction continues, the ocean basin between the two will become smaller and smaller until Westcontinent and the volcano collide. Also the more Westcontinent and the volcano move together, the more oceanic crust is subducted and destroyed. These ocean basins, which are disappearing in a subduction zone, are called **remnant oceans.**

STAGE 6—ISLAND ARC COLLISION MOUNTAIN BUILDING

LATE VOLCANIC MOUNTAIN BUILDING   Westcontinent and the volcanic island have moved closer together, and the remnant ocean basin is now much smaller. The Eastcontinent has also come onto the cross section, but it is still far away and has no influence at this stage. The ongoing subduction and fractional melting of the oceanic crust has built a volcano perhaps 7–8 km off the ocean floor, penetrating the ocean surface to form an island The volcano's center (**mobile core**) is made of many diorite, and probably a few granitic, batholiths. All of this has set in motion several more processes.

Sedimentary Processes   As soon as the volcano breaks the surface, weathering and erosion processes attack it and form lithic sediments which wash into the sea on all sides. Sediments on the backarc side just spill onto the ocean floor as turbidity currents[1] and stay there undisturbed. On the forearc side, however, the sediments pour into the trench. A trench is like the mouth of a conveyor belt and sediments do not stay there long. Instead they are scraped off the subducting oceanic crust into a melange deposit, or they are partially subducted and metamorphosed. A **melange** is a chaotic mixture of folded, sheared, and faulted blocks of rock formed in a subduction zone.

It is also normal, if the climate is right, for reefs to grow around the island. These limestones typically interbed with the coarse-grained lithic breccias and conglomerates eroding from the volcano and the volcanic sands on the beach. During a volcanic eruption, then, lavas may interbed with the limestone and lithic breccia to form a very unusual association of rocks.

---

[1]A turbidity current is an underwater avalanche. A swirling mixture of sediment and water flows rapidly down underwater slopes and spreads across the basin floor. The turbidity current slows down gradually, and the resulting turbidite deposit is a graded bed, coarse sediment on the bottom and getting ever finer toward the top.

Paired Metamorphism Two major kinds of metamorphism are common in a volcanic arc. The first is regional metamorphism (low to high temperature, and medium pressure) formed inside the volcano by heat from the batholiths and the intense folding and shearing occurring there. The metamorphism produces greenschist, amphibolite, and granulite facies rocks, which, because the parent rocks are mafic oceanic crust, are chlorite- and epidote-rich (greenschist), amphibole-rich (amphibolite), and pyroxene-rich (granulite) rocks. Also earlier, now crystallized, intermediate and felsic batholiths may be converted into gneisses and migmatites.

The second metamorphism is high pressure–low temperature **blueschist metamorphism** formed in the melange of the trench. It is high pressure because this is a convergent boundary and the trench sediments are being rapidly subducted between two plates. The low temperature is because cool surface rocks are rapidly subducted and do not have time to heat up. These two belts of metamorphism form a **paired metamorphic belt,** which is always the result of subduction.

Ancient and modern volcanic island arcs are very common. Modern examples are Japan, the Aleutian Islands of Alaska, and the Malaysian archipelago, including the islands of Java, Borneo, and Sumatra. Ancient examples are not as obvious because they eventually collide with another island arc or a continent and are hidden, but that is the next step in the model.[1]

STAGE 7—ISLAND ARC–CONTINENT COLLISION AND MOUNTAIN BUILDING

The remnant ocean basin has closed and the volcanic island arc has collided with the divergent continental margin creating a large mountain. Collision mountain building is of two basic kinds: (1) island arc–continent collision and (2) continent–continent collision.[2] The island arc–continent collision is described here, the continent–continent collision later.

Collision Orogenies

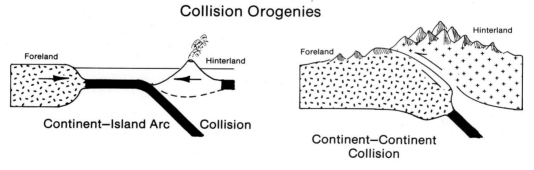

Continent–Island Arc Collision

Continent–Continent Collision

Observe the geometry in the stage 7 cross section. Because the subduction zone dips east, the island arc has attempted to slide up over the edge of the (formerly divergent, but now) convergent continental margin. We can generalize this: in every collision orogeny one plate is going to ride up onto the edge of the other. The overriding plate is called a **hinterland.** The overridden plate is called a **foreland.** It does not matter what is on the edge of the plate (volcanic arc, hot spot volcano, continent), or which way the subduction zone dips; the overriding piece is always the hinterland, the overridden piece is always the foreland.

Suture Zone During the collision the first part of the volcanic arc to be affected is the trench

---

[1]Volcanic island arcs caused by subduction should not be confused with hot spot volcanoes such as those in Hawaii, the Virgin Islands, the Azores, and Bermuda.

[2]Other collision combinations are possible, such as an island arc–island arc, or an island arc–hot spot volcano. Almost any combination which does not violate earth processes is possible and has most likely occurred somewhere at some time. Here we want to keep it simple.

melange. The melange has been accumulating for a long time as it was scraped from the descending oceanic crust, and now it is thrust up over the hinterland along a major thrust fault where it is smeared out and sheared even more. In the end the melange belt will go from being a hundred or more kilometers wide to maybe only 10 km wide. This narrow zone of ground up, smeared out rock is the **suture zone**, and it is the boundary zone which separates the two plates which have collided and are "sutured" together.

Hinterland Mountain   The volcanic island arc may have been a few kilometers high before the collision, but now it is dramatically thrust up even higher into snow-capped mountain peaks. Along the way very large thrust faults dipping back toward the hinterland carry rock toward the foreland. Behind the major mountain peaks some volcanic activity may continue from the last magmas rising from the subduction zone. It is the last gasp, however, because with the collision subduction stops, volcanic activity stops, mountain building stops, and the only thing remaining is for the mountain to erode.

Foreland   Two significant things happen to the foreland. First, the thick wedge of older divergent continental margin sediments gets compressed, thrust faulted toward the foreland, and folded into anticlines and synclines. Also, the continental margin sediments closest to the island arc batholiths may be mildly regional metamorphosed forming marble, quartzite, slate, and phyllite.

Second, the region inland from the collision sinks into a **foreland basin** which fills with a thick **clastic wedge** of sediments. Foreland basin clastic wedges are common in the geologic record and are commonly called **geosynclines.** Foreland basin deposits are very variable in composition and environment. Because the hinterland mountain is composed of an island arc, the sediments eroded from it are dominantly lithic in composition, with varying amounts of sodic plagioclase feldspar from the intermediate igneous rocks. In the foreland basin, clastic sediments may be deposited in mostly deep-water environments, mostly shallow-water environments, mostly terrestrial environments, or some combination of these. Inland from the foreland basin the clastic wedge thins and becomes finer and finer grained until it merges with cratonic sediments.

In time, the hinterland mountains will erode to sea level (a peneplain). But by that time the hinterland (the former island arc) is permanently sutured to the Westcontinent (stage 8 cross section). Westcontinent has grown larger because of the island arc–continent collision, but this was possible only because subduction and fractionation of oceanic crust created the intermediate and felsic batholiths which compose continental crust.

STAGE 8—CORDILLERAN MOUNTAIN BUILDING

The subduction zone under the stages 5 and 6 island arc is now dead, but the Eastcontinent and Westcontinent are still being driven together by forces outside the cross section. Therefore, another subduction zone has to begin. It could begin anywhere within the ocean basin and form another island arc, and it could dip in any direction. But in this model, decoupling occurs in the east under the edge of the Eastcontinent, forming a cordilleran type of mountain building.

The processes of trench formation, subduction and fractional melting of the oceanic crust, melange deposition, and blueschist metamorphism are the same here as for an island arc orogeny. On the other hand, the rising intermediate to felsic batholithic magmas now inject into the thick wedge of continental margin sediments heating them to very high grade regional metamorphism (amphibolite to granulite facies). If the sediments are limestones and quartz sandstones, the metamorphic rocks will be marbles and quartzites. Less mature sandstones and shales will form slates, phyllites, schists, and gneisses. It is also quite likely that the basement batholiths under the divergent continental margin will be metamorphosed into gneisses and migmatites.

Along with all the heat and metamorphism, the old divergent continental wedge of sediments and batholiths are uplifted along major thrust faults until they form towering mountains. The Andes in South America and the Cascades in Washington, Oregon, and northern California are mountains of this type.

Inland from the volcanic front, in the backarc region, **backarc spreading** occurs. Heat rising from above the subduction zone creates a small convection cell which stretches the continental crust so that normal faults develop into deep graben. Superficially, this may seem like an axial rift, but it forms under very different conditions and processes.[1]

The graben fills with a great complex of deposits, including coarse clastic sediments in alluvial fan and braided rivers, and intermediate to felsic volcanics rising from the subduction zone. Because the sourceland composition is so variable (divergent margin rocks, suture zone rocks, metamorphics, and volcanics and, when erosion is deep enough, felsic and intermediate batholithic rocks), the sediments eroded from it are rich in quartz and (many kinds of) lithics, plus lesser amounts of feldspar (sodic plagioclase and orthoclase).

The volcanics in the backarc basin begin mafic, but slowly turn into intermediate (e.g., andesite) and, finally, felsic (e.g., rhyolite) rocks. In the latter stages, granite batholiths will invade the now mostly filled graben.

STAGE 9—CONTINENT–CONTINENT COLLISION MOUNTAIN BUILDING

This mountain building has many of the same elements as the island arc–continent collision: hinterland, foreland, suture zone, foreland basin, and a towering mountain range. One major difference is that it is now divergent continental margin sediments, or their metamorphic equivalents, which are being thrust toward the foreland. We would, thus, expect the sediments filling the foreland basin to be quite different in composition from those eroding from an island arc, even if they are deposited in very similar depositional environments. These sediments, because of the complex source land, will show the variability found in the back arc basin and be rich in quartz, (many kinds of) lithics, and some feldspar (sodic plagioclase and orthoclase).

STAGE 10—PENEPLAINATION AND STABLE CONTINENTAL CRATON

The cycle which began in stage 1 now comes to an end. The original stable continental craton which was rifted into two pieces is now back together. Note, however, that this new continent is quite complex and the rocks exposed at the surface are very diversified. In addition to the original Westcontinent and Eastcontinent blocks, there are now two foreland basin clastic wedges (probably filled with quite different sediments since, one was eroded from a volcanic arc and one from a cordilleran mountain). There are two suture zones of melange and a host of different igneous and metamorphic rocks. Nonetheless, when everything is finally weathered to completion and the continent is eroded to the peneplain, the simple ideal model for sedimentary rocks will be in force and the continent will be covered by a veneer of quartz sand and limestones (the shale, in time, is washed off the continent edge into the ocean).

[1] Backarc spreading also occurs behind a volcanic island arc. We did not include it at stage 6 for simplicity.

In stage 1 we began with an ideal continent, assuming it was homogenous in structure and composition. In light of the history we have just presented, it should be clear that the original continent was not homogeneous. Over and over and over, since the first crust solidified, the processes of subduction have been making new continental crust. Collisions have been welding them together, and rifting has been fragmenting them.

It is the work of geologists to read great events in the rocks of the earth's crust, but it is also something like a flea trying to understand the great dog it is living on: endlessly fascinating, endlessly frustrating, and immensely satisfying when we glimpse a little of the greatness of it.

A Plate Tectonic Rock Cycle

No rock is accidental. Every rock forms under a specific and unique set of tectonic circumstances. Find the rock and you know what circumstances existed at that spot at the time that rock formed. Change the circumstances and a different rock forms. This is the underlying premise of this laboratory and a basic tenet of the science of geology.

THIS LABORATORY ASSUMES YOU KNOW OR CAN DO THE FOLLOWING:

1. Label on a cross section the following: **forearc, backarc, remnant ocean basin, paired metamorphic belt (blueschist** and **regional), foreland, hinterland, hot spot,** and **suture zone.**

2. Distinguish between island arc–type and cordilleran–type subduction zones.

3. To complete the critical reasoning problems in this laboratory you must have read and understood the plate tectonic rock cycle discussed in the Preliminary preceding this exercise. Some answers cannot just be looked up without understanding the whole model because you will be asked to apply the principles in the model to new situations.

If you have not read the preliminary to a plate tectonic rock cycle, *do it now*. If you do not understand parts of it, read your notes, your textbook, or ask your neighbor.

There are three parts to this laboratory:

PART ONE—PLATE TECTONIC RELATIONSHIPS

PART TWO—IGNEOUS, SEDIMENTARY, AND METAMORPHIC ROCKS IN THE PLATE TECTONIC ROCK CYCLE Where you are to draw on all your knowledge of rocks gained in previous labs.

PART THREE—CRITICAL REASONING PROBLEMS Concerning the meaning and implications of concepts in the plate tectonic rock cycle.

PLATE TECTONIC RELATIONSHIPS

☐ (1) Below is a cross section of a portion of the earth's crust. Label it with the following plate tectonic features.

⇒ Forearc ⇒ Backarc ⇒ Remnant Ocean Basin
⇒ Paired Metamorphic Belt (Blueschist and Regional)
⇒ Foreland ⇒ Hinterland ⇒ Hot Spot ⇒ Suture Zone

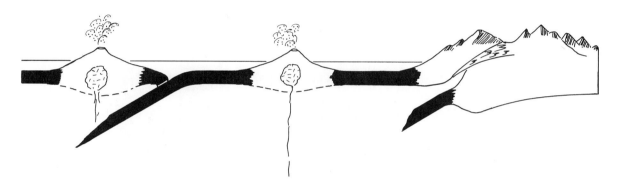

Part Two

IGNEOUS, SEDIMENTARY, AND METAMORPHIC ROCKS IN THE PLATE TECTONIC ROCK CYCLE

☐ (1) Get a tray of igneous, sedimentary, and metamorphic rocks. Work together in pairs.

☐ (2) Cross sections from the plate tectonic rock cycle are reproduced below. In each cross section are one or more arrows, identified by letter, pointing to specific locations in the cross section. Below each cross section is a description of each rock or the processes under which it formed.

STEP A From your collection, find a rock which is most likely to form in the circumstances indicated on the cross section, label it with the letter from the cross section, and place it to one side.

STEP B In the lettered space provided on the next page each cross section, write the name of the rock you believe forms under those circumstances.

STEP C   On a separate piece of paper briefly explain the processes by which each rock formed. It is important to integrate everything you already know: plate tectonic processes and the specific conditions under which each rock forms.

STEP D   Explain to your instructor your identifications and analysis so that you are both sure you understand why no rock is accidental.

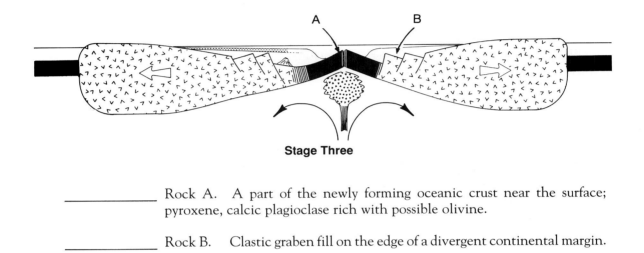

Stage Three

_____   Rock A.   A part of the newly forming oceanic crust near the surface; pyroxene, calcic plagioclase rich with possible olivine.

_____   Rock B.   Clastic graben fill on the edge of a divergent continental margin.

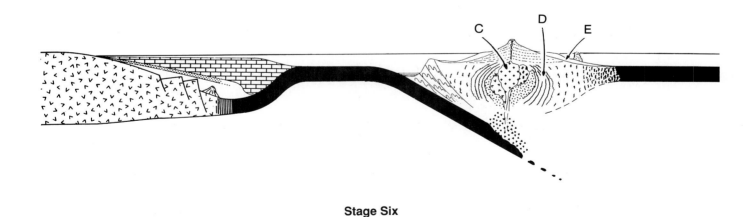

Stage Six

_____   Rock C.   Coarse-grained batholithic rock formed in the mobile core of a volcanic arc.

_____   Rock D.   Medium-grade, regional facies metamorphosed ultramafic igneous rock.

_____   Rock E.   Clastic rocks deposited in the backarc region of a volcanic arc.

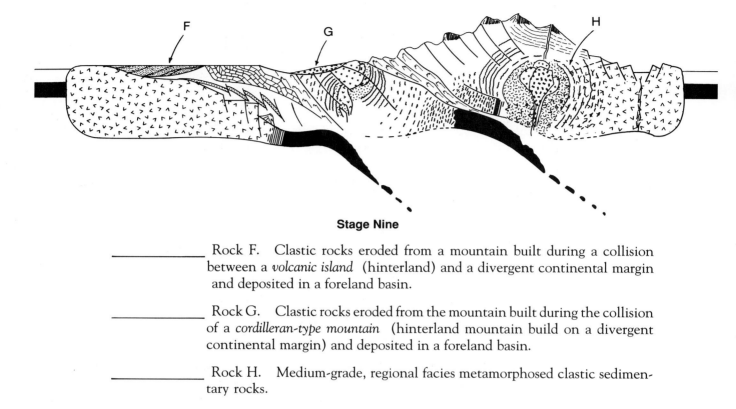

Stage Nine

_____ Rock F.   Clastic rocks eroded from a mountain built during a collision between a *volcanic island* (hinterland) and a divergent continental margin and deposited in a foreland basin.

_____ Rock G.   Clastic rocks eroded from the mountain built during the collision of a *cordilleran-type mountain* (hinterland mountain build on a divergent continental margin) and deposited in a foreland basin.

_____ Rock H.   Medium-grade, regional facies metamorphosed clastic sedimentary rocks.

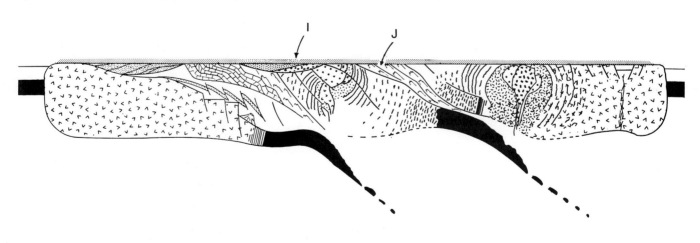

Stage Ten

_____ Rock I.   Clastic rocks deposited on a continent which has been eroded to a stable craton.

_____ Rock J.   Suture zone rocks.

Part Three

CRITICAL REASONING PROBLEMS

☐   (1)   Get the chart titled "A Plate Tectonic Rock Cycle" from the pocket at the back of the manual and lay it open on the table so you can refer to it.

☐   (2)   On a separate piece of paper write answers to the following critical reasoning problems about the plate tectonic rock cycle.

       A.   For each problem there is a statement followed by several possible answers. For each possible answer you must indicate whether you REJECT it or ACCEPT it.

       B.   Logically and factually, in *writing*, explain your acceptance or rejection of *each* and *every* answer. There is NO CREDIT for a "right" answer, only for the analysis.

       C.   For each problem there is one ACCEPT and two REJECTS. Some answers will differ by subtle distinctions, but you must find the *best* answer and explain why the other answers are less correct. It is more important that you be able to explain why the wrong answers are wrong than the right answers are right.

       D.   You may discuss the problem with classmates, but when you write your analysis it must be your own thinking, in your own writing.

Problem Number One

ACCEPT one, REJECT two. From your knowledge of the origin and history of continental and oceanic lithosphere gained from a study of the plate tectonic rock cycle:

_____ 1.   Oceanic crust can exist as long as continental crust.

_____ 2.   For the most part oceanic crust is going to be younger than most continental crust.

_____ 3.   Continental crust is more easily subducted than oceanic crust.

ACCEPT one, REJECT two. Problems Two, Three, and Four each show a cross section of a portion of the earth's crust. Below each cross section are three possible plate tectonic models. You are to ACCEPT the one plate tectonic model most likely to result in the association of rocks seen in the cross section.

Problem Number Two

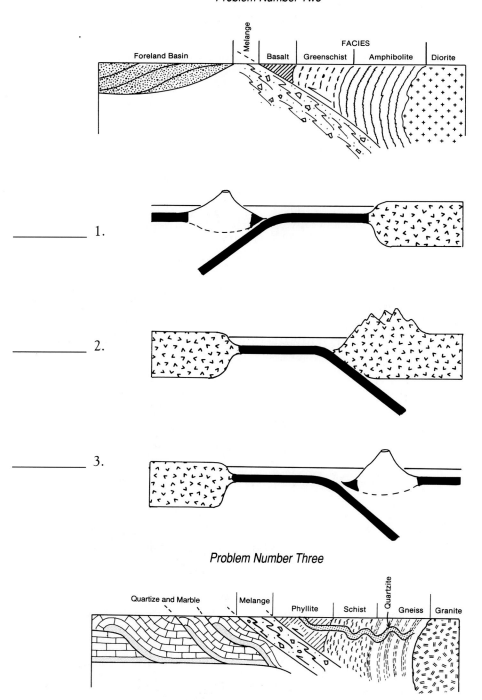

Problem Number Three

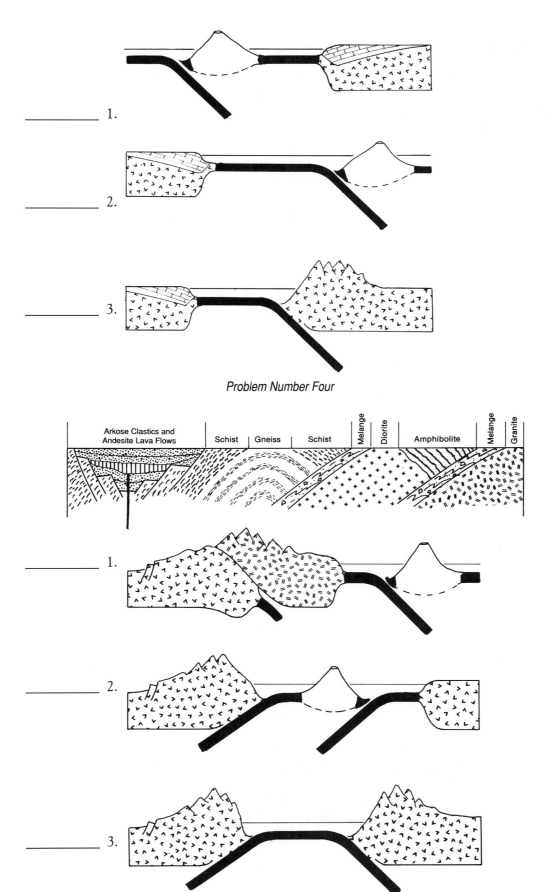

_____ 1.

_____ 2.

_____ 3.

Problem Number Four

| Arkose Clastics and Andesite Lava Flows | Schist | Gneiss | Schist | Melange | Diorite | Amphibolite | Melange | Granite |

_____ 1.

_____ 2.

_____ 3.

215

Problem Number Five

ACCEPT one, REJECT two. During an island arc mountain building, such as in stage 6, the sediments eroded from the volcanic arc are most likely going to be:

_____ 1. Lithic in composition.

_____ 2. Arkosic in composition.

_____ 3. Quartz wacke in composition.

Problem Number Six

ACCEPT one, REJECT two. In the tectonic conditions which exist during stage 1 of the plate tectonic rock cycle, the sediment most likely found on the continent will be:

_____ 1. Quartz sandstone.

_____ 2. Arkose sandstone.

_____ 3. Lithic sandstone.

Problem Number Seven

ACCEPT one, REJECT two. In the laboratory on *Earth Deformation and Geological History* are a number of structural cross sections (pages 145–147). Cross–section 5 (page 164) most likely formed during:

_____ 1. Stage 2.

_____ 2. Stage 4.

_____ 3. Stage 7.

Problem Number Eight

ACCEPT one, REJECT two. In the laboratory on *Earth Deformation and Geological History* are a number of structural cross sections (pages 145–147). Cross–section 6 (page 166) most likely formed during:

_____ 1. Stage 3.

_____ 2. Stage 6.

_____ 3. Stage 7.

INTRODUCING PROBLEM NUMBERS NINE–ELEVEN

For these problems you need to know how to read ternary diagrams explained in the Level Three classification, *Sedimentary Rock* lab, page 91. You do not need to have done the Level Three classifications, you just need to be able to read a ternary diagram.

Below is a quartz-feldspar-lithic (QFL) ternary diagram. On the diagram are plotted a number of fields labeled A, B, C, and D. Each of these fields contains the distribution of sediments which result from a particular kind of tectonic setting. The evidence is empirical; that is, it is observed that the particular kinds of sediment in each field come from the particular tectonic condition. These problems deal with your ability to predict, based on what you know from the plate tectonic rock cycle, the tectonic condition which produce sediments of particular compositions.

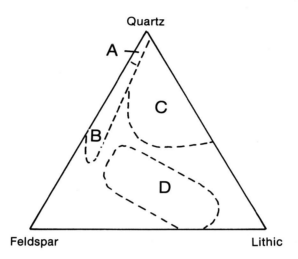

Problem Number Nine

ACCEPT one, REJECT one. A sediment found in field B is most likely:

_____ 1. Lithic and feldspar rich.

_____ 2. Quartz and feldspar rich.

Problem Number Ten

ACCEPT one, REJECT one. A sediment found in field D is most likely:

_____ 1. Quartz and lithic rich.

_____ 2. Feldspar and lithic rich.

Problem Number Eleven

ACCEPT one, REJECT two. A sediment found in field C is likely to:

_____ 1. Have less quartz than field D.

_____ 2. Have less feldspar than field D.

_____ 3. Have less lithics than field D.

Problem Number Twelve

ACCEPT one, REJECT two. During stage 2 the erosion of the horst blocks on either side of the axial rift are most likely going to produce a sediment found in field:

_____ 1. A.

_____ 2. B.

_____ 3. C.

Problem Number Thirteen

ACCEPT one, REJECT two. An island arc orogeny such as in stage 6 is most likely to result in a sediment of which composition field?

_____ 1. B.

_____ 2. C.

_____ 3. D.

Problem Number Fourteen

ACCEPT one, REJECT two. A sediment with a composition found in field C would most likely result from which stage in the plate tectonic rock cycle?

_____ 1. Stage 7.

_____ 2. Stage 8.

_____ 3. Stage 10.

Problem Number Fifteen

Someone who has not taken a geology course says to you, "I don't see how rocks cannot be accidental. Mountains seem to be found in all kinds of places, on the edge of as well as within continents, and volcanoes and earthquakes happen all over the earth."

Write a statement where you explain to this person why no rock is accidental.

INTRODUCING PROBLEM NUMBERS SIXTEEN–EIGHTEEN[1]

What makes a hypothesis scientific?

To be scientific an hypothesis must have three characteristics:

1. It must *explain* with natural laws and processes why a part of the world is as it is.

2. It must make *predictions* about things we don't yet know.

---

[1] These problems are for advanced students or geology majors.

3. It must be *testable*; that is, the predictions it makes must be capable of being demonstrated false. If the prediction does turn out to be false, then the hypothesis must necessarily be false too.[2] A prediction stated in such a way that it takes no risk, that there is no conceivable way to prove it wrong, is unscientific.

These last problems ask you to make scientific predictions about the results of hypothetical, but very plausible, plate tectonic situations—and explain the known processes which support your predictions.

Making predictions is an imaginative and creative process because we must describe what we have not yet seen. There are no "right" predictions to these problems, but there are better predictions. A better prediction is one which provides the best supporting explanations, using known processes, logically argued.

Based on your knowledge of the plate tectonic rock cycle, imagine the processes which would occur in the following three situations. Imagine you want to look for a case like this in the geologic record. To recognize it when you see it, you would need to know what to look for.

☐   1. For each of the processes or terms below, what specific conditions or rocks would you predict will form as a result of the plate tectonic scenario on page 220. Refer to type of rock, its spatial arrangement, associations with other rocks, and so on.

  Would you expect:

  METAMORPHISM: Regional? blueschist? both? Where are they located relative to each other?

  IGNEOUS ACTIVITY: Where is the magma being generated? What kind is it? Is it intrusive, extrusive, both? Does the type of igneous activity change with time?

  SEDIMENTARY ROCKS: Are they likely to be arkosic or lithic or carbonates? What is their maturity? In what kinds of environments are they being deposited? Do the sediments evolve with time and changing tectonic conditions?

  STRUCTURAL GEOLOGY: Are the forces compressional or tensional? What structures result from these forces? What direction are the faults dipping? Do the forces change with time and circumstances?

☐   2. What similar situations are there in the plate tectonic rock cycle that you might confuse your model with? How would you tell one from the other?

☐   3. When you have made your predictions and supporting arguments, compare them with those of other students. Whose predictions are the best argued, and why? Are two different sets of predictions equally valid? Can your prediction be synthesized with someone else's to make a better, more inclusive prediction?

---

[1] If the prediction turns out to be true, then the hypothesis is strengthened but not demonstrated true. No prediction can prove a hypothesis true because the next prediction may turn out to be false.

Problem Number Sixteen

Observe the following scenario of two island arcs, with a remnant ocean between them. They collide, build a mountain, and erode down to a peneplain. Predict the results by, first, drawing a cross section of the geology at the peneplain stage (like those for Problems two, three, and four) and, second, explaining in writing why you drew it as you did.

Problem Number Seventeen

The following scenario is in two stages. In the first stage, there is a divergent continental margin on the right, an ocean basin on the left with a left-dipping subduction zone creating an island arc. The remnant ocean basin closes and the island arc and continent collide and erode to a peneplain. In the second stage, a second, right-dipping subduction forms on the left side of the sutured and eroded island arc/continent couplet. Predict the results by, first, drawing a cross section of the geology at the peneplain stage (like those for Problems two, three, and four) and, second, explaining in writing why you drew it as you did.

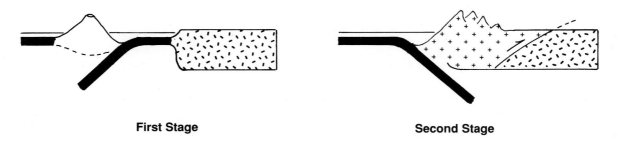

First Stage   **Second Stage**

Problem Number Eighteen

On the right is a continent with a right-dipping subduction zone under its edge creating a cordilleran-type orogeny. On the left is an ocean basin. Within the ocean basin is a divergent boundary. As subduction continues, the continent will eventually override the divergent boundary and subduct it. Predict the results by, first, drawing a cross section of the geology (like those for Problems two, three, and four) and, second, explaining in writing why you drew it as you did.

Topographic Maps

PURPOSE

A topographic map shows the lay of the land; it gives a bird's-eye view of the location of hills and valleys, rivers and streams, the steepness of slopes, and many other features natural and human made. Topographic maps have so many uses for geologists, engineers, farmers, city planners, persons planning a hiking trip, and innumerable others that they are an invaluable tool and worthwhile knowing how to use.

The topographic maps we use are **quadrangle maps** prepared by the United States Geological Survey. A quadrangle is a rectangular area (but not a geometric rectangle) of the earth's surface. Such maps are available for almost any part of the United States. Other countries also publish topographic maps.[1]

To use a topographic map effectively, you need to understand its conventions, how contours reflect topography, and the large amounts of information printed on it. The purpose of this lab is to:

1. Introduce the topographic map scales commonly in use, the longitude and latitude system for locating places (and topographic maps) on the earth, and most of the many other kinds of information printed on a topographic map.

2. Develop skills in reading contour lines and using them to understand landforms.

There are five parts to this laboratory:

PART ONE—READING TOPOGRAPHIC QUADRANGLE MAPS

PART TWO—SEEING THE LAY OF THE LAND WITH CONTOUR LINES AND TOPOGRAPHIC PROFILES

PART THREE—THE PUBLIC LAND SURVEY SYSTEM: TOWNSHIP, RANGE, AND SECTION

PART FOUR—RELIEF AND GRADIENT

PART FIVE—AN INTRODUCTION TO LANDFORMS

---

[1] See end of exercises for places to order United States Topographic maps.

Part One
READING TOPOGRAPHIC QUADRANGLE MAPS

MAP SCALES AND MAP DISTANCES

A topographic map is a scale model representing a part of the earth's surface. Topographic maps usually cover two degrees of arc or less. On most topographic maps, the area covered is much less than a degree and is measured in minutes and seconds of arc. (There are 60 seconds in a minute and 60 minutes in a degree, written, for example, as 75° 35' 15": seventy-five degrees, thirty-five minutes, and fifteen seconds.) The scale of the map is how much smaller the map is than the real world it represents and is expressed in three different ways.

A **fractional scale** expresses how much smaller the map is as a fraction, such as 1/24,000 or 1:24,000. This means that the map is 1/24,000 the size of the real world. Any unit distance on the map is 24,000 times larger in reality. For example, 1 centimeter (inch, foot) on the map equals 24,000 centimeters (inches, feet) on the ground. Most topographic maps are in standard sizes, such as 1:24,000, 1:62,500, and 1:250,000, although many scales of maps have been published.

A **verbal scale** converts the map scale into terms which have more meaning for us, such as 1 inch on the map equals 2,000 feet on the ground (or 1 inch on the map = 24,000 inches on the ground divided by 12 inches per foot = 2,000 feet). A verbal scale may not be precise. For example, 1:62,500 is only approximately 1 mile (1 inch = 1 mile because 1 mile is 63,360 inches), but this is close enough for rough estimates.

A **graphic scale** is printed in the lower margin of the map showing miles, feet, and kilometers so that, with calipers or a ruler, distances on the map can easily be scaled out and, when the map is enlarged or reduced, still gives a reliable scale.

☐ (1) Your instructor will provide you with one or more topographic quadrangle maps. Answer the questions below.

☐ (2) Please do not write on the maps.

☐ (3) Some conversion factors you may need are

1 mile =	1 kilometer =	1 foot=
5,280 feet	1,000 meters	.305 meters
63,360 inches	.62 mile	
1.6 kilometers		

1 yard =	1 meter =
36 inches	100 centimeters
.914 meters	39.37 inches

1. Calculate the verbal scale in number of feet per inch, miles per inch, and kilometers per centimeter for each of the following fractional map scales.

1:24,000	1:62,500	1:250,000
_____ ft/in.	_____ ft/in.	_____ ft/in.
_____ mi/in.	_____ mi/in.	_____ mi/in.
_____ km/cm	_____ km/cm	_____ km/cm

_____ 2. Write the **fractional scale** of the map provided by your instructor (look at bottom, center).

3. Write the **verbal scale** of the map provided by your instructor; how many feet and miles per inch and kilometers per centimeter does the map represent?

_____ ft/in.

_____ mi/in.

_____ km/cm

_____ 4. Measure the verbal scale of feet per inch you wrote for 3 above against the graphic scale at the bottom of the map. Are they indeed the same? If not, there is a mistake somewhere.

LATITUDE AND LONGITUDE

Every topographic map can be located precisely on the earth's surface by the international grid system of latitude and longitude (illustration, next page). Lines of **latitude** circle the earth parallel to the equator. The equator is zero degrees latitude, and latitude increases degree by degree in each direction north and south of the equator toward the poles. Maximum latitude is 90° at each pole.

Lines of **longitude** circle the earth from pole to pole. The first, or zero, line of longitude is the **prime meridian**, which runs through Greenwich, England (near London). The British established an observatory here to serve as a reference point for their navy and merchant fleets, and it has become the international standard. Longitude increases east and west of Greenwich degree by degree around the globe until 180° is reached on the opposite side of the globe. With the longitude and latitude system, every place on the earth is either east or west of Greenwich, and north or south of the equator. All North American latitudes are north of the equator, all longitudes are west of the prime meridian, but you should get in the habit of writing north or south, east or west after latitude and longitude numbers. The longitude and latitude of every topographic map is shown at the corners of the map, with subdivisions shown along the map edges.

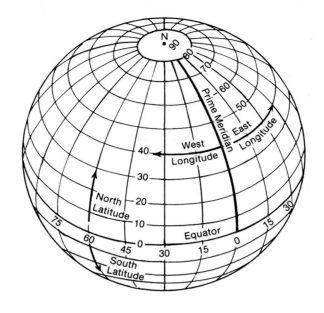

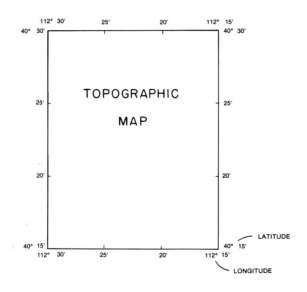

☐   (4)   Answer the following questions about latitude and longitude for your map(s).

_____ 5.   Write the latitude for the north end (top) of your map.

_____ 6.   Write the latitude for the south end (bottom) of your map.

_____ 7.   What is the distance in degrees, minutes, and seconds from north to south on this map? (line 5 minus line 6)

_____   8.   Write the longitude for the west (left) side of the map.

_____   9.   Write the longitude for the east (right) side of the map.

_____   10.   What is the distance in degrees, minutes, and seconds from west to east on this map? (line 8 minus line 9)

_____   11.   Are the distances in degrees, minutes, and seconds for your map the same for longitude and latitude? It is not always this way.

_____   12.   Record the map series printed in the upper left of the map.

E _____   13.   With a ruler measure to the nearest millimeter (or 32nd of an inch) the

W _____   north-to-south height of the map **on both the east and the west sides.**

_____   14.   Are they the same height?

ft _____   15.   Calculate the real distance from north to south in feet; in kilometers.

km _____

N _____   16.   With a ruler, measure to the nearest millimeter (or 32nd of an inch) the east-

S _____   to-west width of the map at **both the north and south ends of the map.**

_____   17.   Are they the same width? If not, why not?

ft _____   18.   Calculate the real distance from east to west at the south end of the map

km _____   in feet; in kilometers.

mi² _____   19.   Calculate the approximate area of the map in square miles and kilometers.

km² _____

_____   20.   Write the exact latitudinal center of the map.

_____   21.   Write the exact longitudinal center of the map.

_____   22.   Choose a landmark on the map and write its longitude and latitude.

OTHER INFORMATION ON A TOPOGRAPHIC MAP

☐   (5)   A topographic map provides a lot of information around the margins. Find and record the following for your map.

_____   23.   What is the title of the map?

Date _____   24.   When was the map published? When was it surveyed? Was it revised and when?

Sur _____

Rev _____

_____ 25. Photo revisions of quadrangle maps are done in purple. Is there any photo revision on your map? When was this revision done?

_____ 26. What organization did the actual mapping of the area?

_____ 27. What map is adjacent to the east?

_____ 28. What map is adjacent to the northwest?

_____ 29. What is the contour interval (contours are discussed below).

A page of topographic map symbols used by the United States Geological Survey is in the pocket at the back of the manual. Get it out and see if your map has the following features. If the symbol is present, sketch it in the space. If the symbol is not found on your map, indicate "not present."

_____ 30. What kinds of highways are represented by the map? symbol color?

_____ 31. Railroad tracks present?

_____ 32. Buildings, churches, schools?

_____ 33. Open pit, mine, or quarry?

_____ 34. Marshes? symbol color?

_____ 35. Woods or brushwood? symbol color?

_____ 36. Benchmarks showing an elevation?

_____ 37. A stream? Is it intermittent? How do you know?

_____ 38. County lines, parish, or municipal boundaries?

_____ 39. Wells?

_____ 40. Power transmission line?

MAGNETIC DECLINATION   The axis of the earth's rotation marks the true North and South Poles. Lines of longitude converge toward the poles and, in the Northern Hemisphere, are closer together at the top of a map than the bottom, as your answers to questions 16 and 17 demonstrate. The sides of a quadrangle map (that is, the longitude lines) always point to true north. North on a topographic map is true north.

A compass points to the **magnetic pole,** which is not at the same location as the axis of rotation. A simple compass will point to both true north and magnetic north only along the two lines of longitude which happen to pass through both (see illustration, page 227). For every other longitude there is a **declination** (an angle) between true north and magnetic north. The declination of every topographic map is indicated at the bottom of the map; it tells you how many degrees east or west true north is from magnetic north. More sophisticated compasses allow you to make an adjustment so that when the needle says north, the axis of the compass is in fact pointing true north. Every time you move to another map the declination of the compass must be adjusted for the map area. This can be important if you need to locate things very precisely.

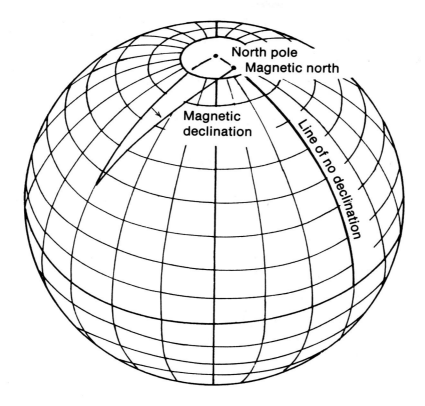

_____   41.   Find the declination indicator on your map. What is the declination?

_____   42.   In what year was the declination measured?

The magnetic north pole does not stay in the same location but slowly moves; therefore, declinations are likely to be less accurate the longer ago they were measured.

Part Two

SEEING THE LAY OF THE LAND WITH CONTOUR LINES AND TOPOGRAPHIC PROFILES

Contour lines connect points of equal elevation. A contour line is formed by the intersection of an imaginary horizontal plane and the surface of the earth. A map showing elevation with contour lines is a topographic map. Contour lines are imaginary, but if you could see one, it would look like a line running horizontally around a hill or along the side of a valley.

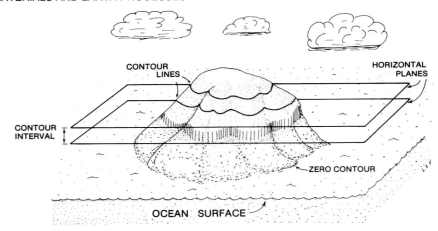

At a lake or the ocean, the shore line is a contour line. Sea level is the first contour line at zero feet above or below mean sea level. All other contour lines are at some selected **contour interval** above mean sea level, for example, 20 ft, 40 ft, 60 ft, 80 ft, and so on. When seen on a map the contour lines seem to circle around hills and point up valleys. The topographic map below has a scatter of points with their elevation written beside them and contour lines interpolated at 20-ft intervals.

TOPOGRAPHIC MAP

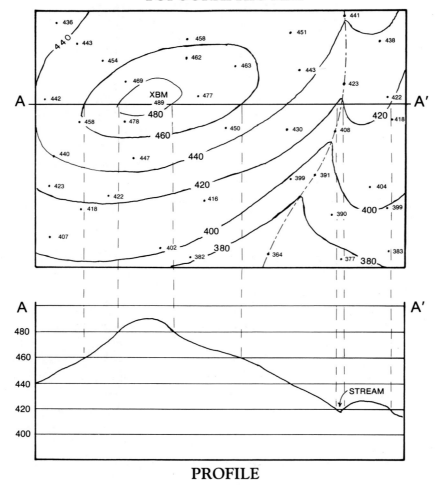

PROFILE

TOPOGRAPHIC PROFILES   A topographic map is a bird's-eye view of the landscape, but that is not the way we on the ground normally see the countryside. We see things in profile, looking at the side of a hill or down a valley. We can reconstruct a countryside's profile from a topographic map. In the example on page 228 a profile has been drawn along line AA', up the side of a hill, over the top, down the other side into a stream valley, and then over a small knoll. This is done by projecting the elevation of a contour line which intersects the line AA' down to the same elevation on the cross section. The profile is then drawn by connecting the dots. In the next few exercises you will learn to use contour lines and to draw topographic maps and profiles.

Some Rules of Contour Lines on Quadrangle Maps

1. **Color**   Contour lines are brown on a published map. **Index contours** are heavier brown lines and designate major units of elevation (usually every fifth contour). The elevations of contour lines are printed at places along the lines.

2. **Slope steepness**   When contour lines are close together, the slope is steep. When contour lines are far apart, the slope is gradual.

3. **Contour shape**   When contour lines outline a hill, they usually have a rounded, smooth shape. Contour lines which trace up a valley and cross a stream are "V"-shaped. In other words, the shape of contour lines is the shape of the feature they represent. Look at the topographic map provided by your instructor, or the St. Paul Quadrangle at the back of the manual, and notice that "V"s and rounded contour lines alternate with each other from hill, to valley, to hill.

4. **Closing contours**   Contour lines eventually close, or connect end to end, creating an enclosed area, although this may happen outside the map area.

5. **Contour lines never branch or split.**

6. **Contour lines never cross,** for this would mean a lower elevation rises above a higher elevation, or vice versa. There is one exception: contour lines for an overhanging cliff will cross lower contour lines; the lower contour lines are dashed to indicate the overhang.

7. **Depressions**   Enclosed depressions or hollows are shown by using hachure lines which point into the depression. See the Mammoth Cave or Jackson Quadrangles at the back of the manual.

☐   (1)   On page 230 is an incomplete topographic map showing only spot elevations and streams. Work alone as you draw the contour lines to create a topographic map.

STEP ONE   Find the highest elevations on the map and mark them with an X. These will be the tops of hills, and contour lines will have to circle them. But remember, when contour lines approach a stream, they will "V."

STEP TWO   Draw contour lines. The lake shore is the first contour line at 600 ft. Use a contour interval of 20 ft to draw the 620 ft, 640 ft, and so on contour lines around the hills. Since spot elevations are rarely on a contour line, you will have to interpolate between elevations. for example, an elevation of 620 ft. Estimate where between 600 and 633 ft the 620 spot would go and mark it. Continue this way spotting 620 elevations between higher and lower elevations. When there are sufficient points, connect them to create the contour line.

STEP THREE   Draw a topographic profile. Use a straight edge to draw line AB. Next, use the straight edge to project the intersection of each contour line, stream, and hilltop which crosses line AB down onto the correct elevation on the **topographic profile.** Complete the elevation by connecting all the points of elevation on the profile.

☐   (2)   Ask your instructor to check your completed map and profile.

TOPOGRAPHIC MAP

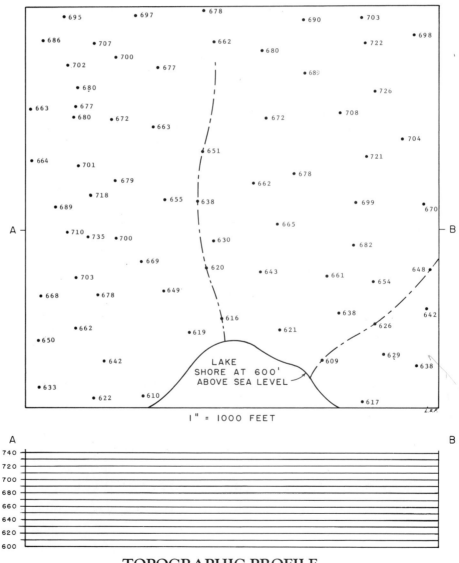

1" = 1000 FEET

TOPOGRAPHIC PROFILE
(1" = 100 FEET)

VERTICAL EXAGGERATION

Standing next to a mountain, it looks very big, and standing next to the Grand Canyon, it looks very deep, at least compared to ourselves. These impressions of height and depth are deceiving. For example, if we take the distance from New York City to San Francisco as 3 feet and draw on that line the highest mountain in North America (about 3 miles high), it would appear only .012 (12/1000) of an inch high. Such a cross section would barely be visible.

To see the mountain on a profile you have to magnify the vertical scale of the line so that hills and valleys show up better. The vertical and horizontal scales are now not the same. The vertical scale is exaggerated. Vertical exaggeration is normal on most profiles and cross sections and will vary depending on the vertical and horizontal scales chosen.

Vertical exaggeration may be calculated a number of ways. One way is if we know how many feet are in an inch of horizontal scale and in an inch of vertical scale on the same profile. If 1 horizontal inch equals 300 feet and 1 vertical inch equals 30 feet, then the vertical exaggeration equals

$$\text{Vertical Exaggeration} = \frac{\text{Horizontal Scale}}{\text{Vertical Scale}} = \frac{300}{30} = 10\textbf{X}\text{ Exaggeration}$$

The fragment of the Soda Canyon Quadrangle, Colorado, on page 238 is a 15-minute series map where 1 inch on the map equals 62,500 inches on the ground, or 5,208 feet per inch. The profiles below the map have a 450-ft-per-inch vertical scale.

$$\text{Vertical Exaggeration} = \frac{\text{Horizontal Scale}}{\text{Vertical Scale}} = \frac{5,208}{450} = 11.5\textbf{X}\text{ Exaggeration}$$

Vertical exaggeration can be calculated from fractional horizontal and vertical scales, too. The 15-minute series Soda Canyon Quadrangle's scale on page 238 is 1/62,500. The vertical scale is 1 inch = 450 feet, or 5,400 inches, or 1/5,400. Vertical exaggeration is:

$$\frac{\text{Fractional Vertical Scale}}{\text{Fractional Horizontal Scale}} = \frac{1/5,400}{1/62,500} = \frac{.000185}{.000016} = 11.5\textbf{X}\text{ Vertical Exaggeration}$$

or

$$\frac{1/5,400}{1/62,500} = \frac{1}{5,400} \times \frac{62,500}{1} = \frac{62,500}{5,400} = 11.5\textbf{X}\text{ Vertical Exaggeration}$$

☐ (3) Calculate the vertical exaggeration. Choose either the Harrisburg Quadrangle on page 239 or the Chief Mountain Quadrangle on page 240.

_____ 43. Vertical exaggeration of Harrisburg profile?

_____ 44. Vertical exaggeration of Chief Mountain profile?

Part Three
THE PUBLIC LAND SURVEY SYSTEM: TOWNSHIP, RANGE, AND SECTION

The **township and range grid system** was adopted and first used in the United States in 1785. It is now in general use in most of the United States, except the original 13 colonies, a few states derived from them, and some parts of the Southwest. The system allows the precise location of small areas with as much precision as one wants. It has also been used to lay out roads and property boundaries as you may have noticed if you have flown over an area where the roads are laid out in a square, checkerboard pattern.

The township and range system is made up independently for each state which has it. The system begins at an **initial point,** often the geographical center of the state. Running through the initial point from east to west is a **baseline**. Running through the initial point from north to south is a **principal meridian.** These two lines establish the grid system. Beginning at the baseline parallel lines are drawn every 6 miles north and south to the state boundary. These 6-mile units are **townships** and are designated township 1, 2, 3, north or south of the baseline (e.g., T1N, T3S). Likewise, lines are drawn parallel to the principal meridian every 6 miles east and west. These 6-mile units are **ranges** and are designated range 1, range 2, range 3, and so on east or west of the principal meridian (e.g., R1W, R2E). The result is a grid of squares of land 6 miles on a side, each of which has a unique township and range number.

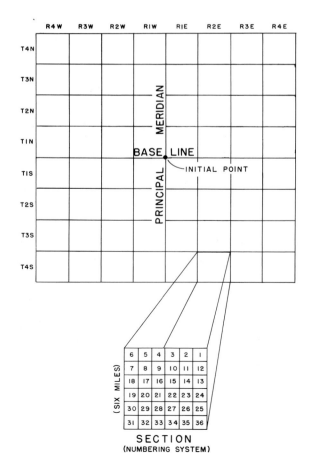

SECTION
(NUMBERING SYSTEM)

For closer location, each 6-mile square of township and range is further subdivided into **sections**. Each section is 1 mile square, so each township and range has 36 sections. Each section is numbered, beginning in the upper right-hand corner and continuing as in the illustration, page 232. You should learn this numbering system since some topographic maps may not have the sections numbered.

SUBDIVIDING THE SECTION   For even more precise location, a section can be subdivided in the following manner.

```
+----------+----------+---------------------+
|          |          |                     |
|   NW¼    |   NE¼    |                     |
|          |          |        NE¼          |
+----------+----------+                     |
|          |::::::::::|                     |
|   SW¼    |:: SE¼ ::|                     |
|          |::::::::::|                     |
+----------+----------+---------------------+
|                     |                     |
|                     |                     |
|        SW¼          |        SE¼          |
|                     |                     |
|                     |                     |
|                     |                     |
+---------------------+---------------------+
```

A section can be divided into four equal quarters: northwest one quarter (NW 1/4), northeast one quarter (NE 1/4), southeast one quarter (SE 1/4), southwest one quarter (SW 1/4). Each one quarter may also be subdivided into quarters, and each of those quarters can be divided into quarters, as many times as necessary to precisely and accurately locate a spot.

A spot is located by describing which quarter(s) it is in —*always beginning with the smallest quarter needed to define the area.* For example, the shaded area in the diagram would be the SE 1/4 of the NW 1/4, of whatever section, township, and range in which it happened to be.

OBSERVE: The grid system of adjacent townships and ranges do not always line up with each other. This may occur because a state line has been crossed going from one T–R system to another, or there is more than one initial point, or adjustments have to be made because the perfectly square grid systems get squeezed as meridian lines converge northward.

☐   (1)   At the back of the manual is a collection of topographic maps, starting on page 281. The first map in the series is the St. David Quadrangle, Arizona, which has a township and range grid system. Answer the following questions.

_____   45.   Name the feature located in the SW 1/4 of Section 32, T. 17 S., R. 23 E.

_____   46.   Write the elevation of the benchmark (BM) located in the NE 1/4 of the NE 1/4 of Section 4, T. 17 S., R. 22 E.

_____   47.   Write the location of Knob Hill (near the center of the map) by township, range, and section.

_____ 48. Write the location of Knob Hill to the nearest quarter of the section as you can.

_____ 49. What is the distance in kilometers from the Prospect in the SW 1/4 of Section 8, T. 17 S., R. 23 E. to the Water at the NW 1/4 of Section 1, T. 18 S., R. 22 E.?

┌ *Part Four* ─────────────────────────────
└────── RELIEF AND GRADIENT ──────

Local relief is the difference in elevation between two points. **Total relief** is the difference between the highest elevation and lowest elevation in a region or on a map.

Gradient is the rate of ascent or descent of an inclined surface of the earth, that is, the steepness of a slope. It is expressed in several ways, but we will use only the fractional method (such as feet of fall per mile) and ratio method (such as how many units of drop for units of horizontal distance). Gradient is frequently calculated for streams but can be calculated for roads or slopes (like for ski slopes), or any inclined surface.

☐ (1) If possible your instructor will provide you with a full size topographic map and the information needed to answer questions 50–58 for that map. If a full-size map is not available, questions 50–58 are for the portion of the Wade Quadrangle included among the topographic maps at the back of the manual.

OBSERVE: If you are calculating the gradient of a meandering river, or of a winding road, you will need to use the actual stream distance or road mileage rather than the straight line distance between the high and low elevations. For the Wade Quadrangle we are calculating the gradient of the slope and so can measure straight down the dip of the slope.

OBSERVE: The gradient between any two points is an average. Along the way gradients can vary from steep to gradual and back again, as when a waterfall gives way to a stretch of calm water. Depending on the purpose for which the gradient is being calculated, these variations may have to be taken into account.

_____ 50. On the Wade Quadrangle locate sections 4 and 8 in the southern portion of the map. In the SE 1/4 of the NW 1/4 of section 4, you will see printed the elevation of a high point; write that elevation on the line.

_____ 51. In the NW 1/4 of the NW 1/4 of section 8, there is a place called Wade with a railroad line running past it. Determine its elevation and write it on the line. Note that the elevation of the railroad line next to Wade does not change for several miles as it travels to the southeast.

_____ 52. What is the difference in elevation between the high elevation in section 4 and the low elevation at Wade and the railroad line in section 8?

_____ 53. Using the bar scale printed on page 279 for 7 1/2 minute maps, determine the distance in miles *straight down the slope* from the high elevation to the railroad line.

_____ 54. Calculate the average gradient of the slope in a fractional scale of feet of fall per mile and write it on the line.

OBSERVE: Your calculated gradient should be in the smallest possible terms, such as 25 feet per mile, not as 100 feet in four miles.

_____ 55. Calculate the gradient as a ratio. A ratio gradient is like one foot of fall for ten feet of horizontal distance (1/10), or one mile of fall for twelve miles of horizontal distance (1/12). Note that the vertical and horizontal scales must be in the same units for a ratio gradient. You can calculate ratio gradients by converting everything to the same units and dividing the distance of fall into the horizontal distance. For example, if the vertical distance is 528 feet, and the horizontal distance is one mile, the ratio is 1/10.

$$\begin{array}{l} \text{Vertical Distance in Feet} = \\ \text{Horizontal Feet in One Mile} = \end{array} \frac{528}{5280} = \frac{1}{10}$$

_____ 56. Write the highest elevation on the map.

_____ 57. Write the lowest elevation on the map.

_____ 58. Calculate the total relief of the map (line 56 minus line 57).

Part Five

AN INTRODUCTION TO LANDFORMS

Topographic maps represent on a flat piece of paper a three-dimensional land surface. You must learn to image quickly and accurately what a land surface is like from just viewing the map.

On pages 237–241 are a number of topographic maps representing widely different areas of North America from deserts, to alpine mountains, to mountain country in eastern North America. By the end of this exercise, you should be able to look at any one of them and describe the topography.

☐ (1) Work alone on this exercise.

☐ (2) Examine the topography along the line of profile for the maps on pages 237–241 Try to visualize the hills and valleys, In the Freehand Profile space sketch in, by eye, what you see as the topography. Do this carefully, but without a straightedge or measurement.

☐ (3) Draw an exact profile for each topographic map on pages 237–241. Find the **index contour lines** (a heavier line indicating large contour intervals) and use a straightedge to transfer these elevations to the appropriate line of elevation on the profile. When you have sufficient points, connect them and draw an exact profile.

☐   (4)   Compare your freehand and exact profiles. If they do not look much alike, there is something wrong and you should find out what. Call your instructor. If your freehand and exact profiles are reasonably alike, congratulations; you understand contours and topographic maps fairly well.

LANDFORM INTERPRETATION

Landforms vary markedly on different parts of the earth. Drawing a profile is the easy task; explaining why the profile looks as it does is not always so easy. We will explore landform interpretation in more detail later, but for the moment let us begin with some simple questions about the topographic maps and profiles you have just drawn.

☐   (5)   Each of the features listed below is found on one of the regions represented by the maps on pages 237–241. Identify the map which has the feature.

    _____ V-shaped valleys.

    _____ U-shaped valleys.

    _____ Alternating ridges and valleys.

    _____ Broad, flat, high areas—plateaus.

    _____ Circular depressions—sinkholes.

    _____ Long gradual slopes.

☐   (6)   Go back and look again at the area represented by each map. Can you hypothesize what the geology is like in each of these regions? Are the rocks flat lying, folded, faulted? Is there any evidence that deposition is going on in the region?

    _____ Area underlain by horizontal strata.

    _____ Area underlain by deformed strata.

    _____ Desert region with extensive sedimentary deposition.

    _____ Region with little rainfall and erosion.

    _____ Area underlain by easily dissolved limestone.

PLACES TO ORDER TOPOGRAPHIC MAPS

Topographic maps, as well as many other maps, are available from many suppliers. In your local area check with outfitter stores, the forest service, hunting stores, large bookstores, or stationery stores. Also check the yellow pages for likely suppliers. Maps are also available by mail.

Map Distribution
U.S. Geological Survey
Box 25286 Federal Center
Denver, Colorado 80225
or
1200 South Eads Street
Arlington, Virginia 22202

GEOSCIENCE RESOURCES
2990 Anthony Road
P.O. Box 2096
Burlington, North Carolina 27216

ANTELOPE PEAK QUADRANGLE, ARIZONA
15 Minute Series

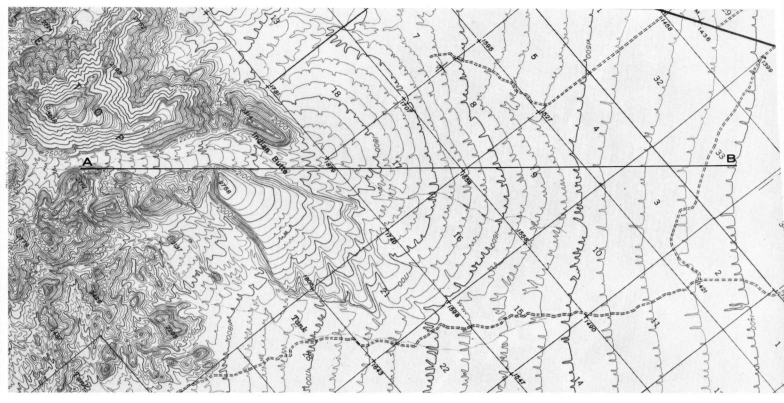

Contour Interval 25′

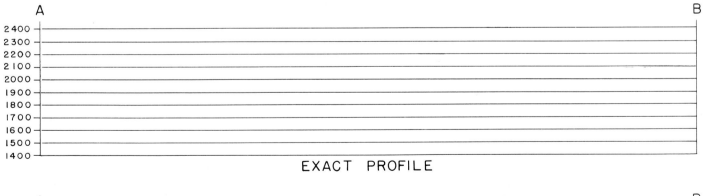

EXACT PROFILE

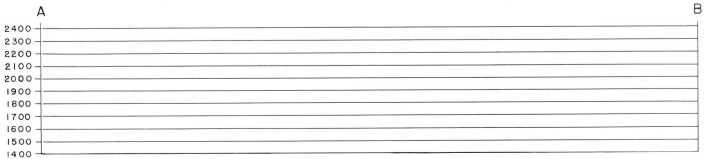

FREE HAND PROFILE

SODA CANYON QUADRANGLE, COLORADO
15 Minute Series

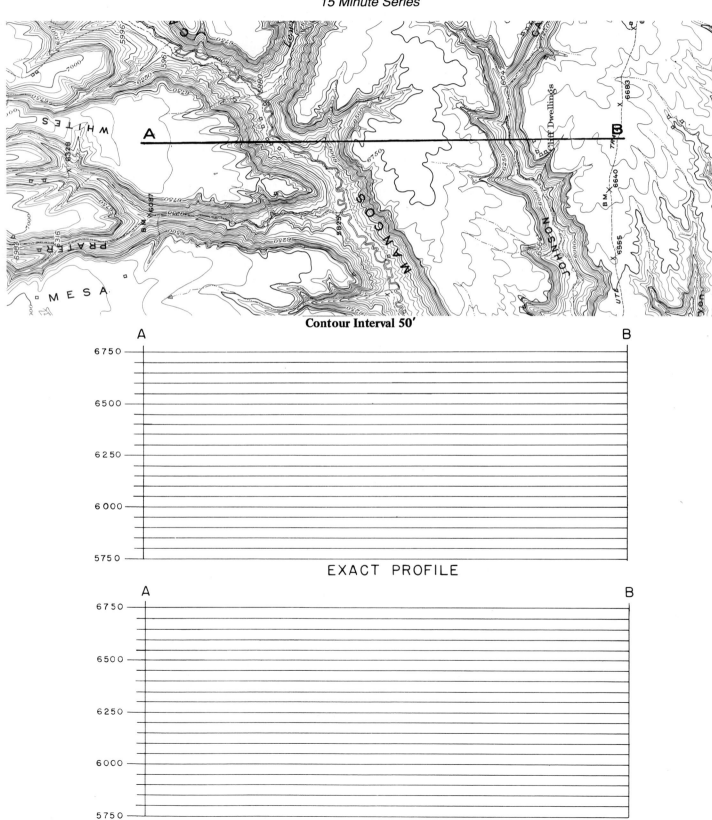

Contour Interval 50′

EXACT PROFILE

FREE HAND PROFILE

HARRISBURG QUADRANGLE, PENNSYLVANIA
15 Minute Series

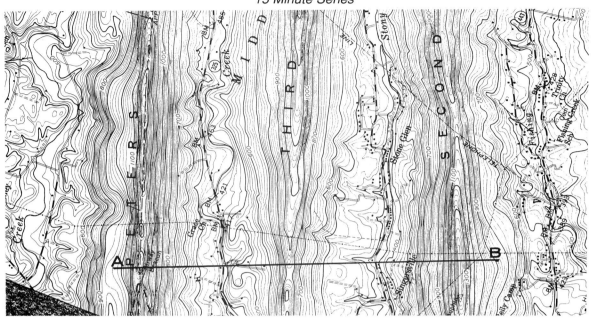

Contour Interval 25′

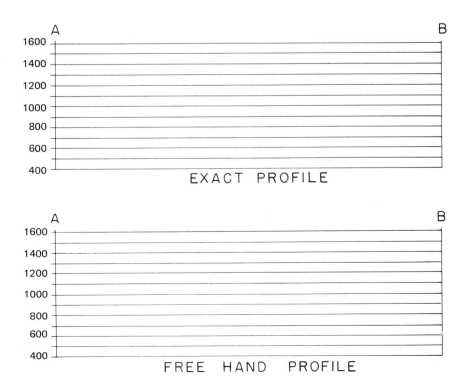

CHIEF MOUNTAIN QUADRANGLE, MONTANA
30 Minute Series

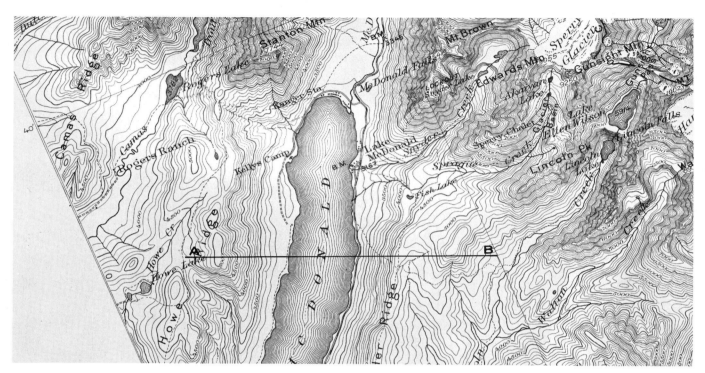

Contour Interval 100′

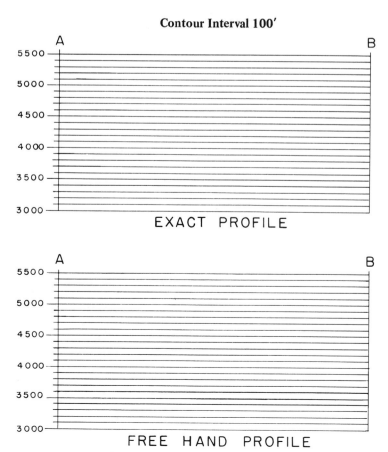

EXACT PROFILE

FREE HAND PROFILE

INTERLACHEN QUADRANGLE, FLORIDA
15 Minute Series

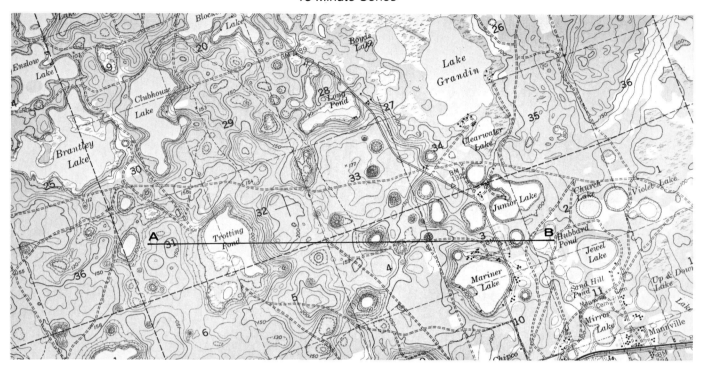

Contour Interval 10′

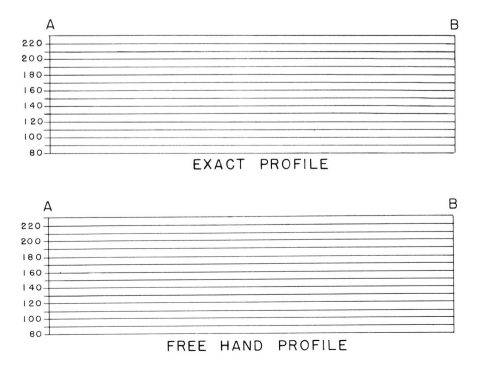

EXACT PROFILE

FREE HAND PROFILE

The Classic Patterns of Landform Development

PURPOSE

These investigations continue the study of topographic maps. Here, however, we will be dealing with surface processes and the development of specific landforms in areas with different structural and climatic conditions. Topographic maps are useful for comparing the development of landforms since they allow comparison of areas normally widely separated. Visualize this exercise as if you were being taken on an aerial trip to the area represented by each of the maps. You are there, looking down on the topography of each area.

Part One

LANDFORM DEVELOPMENT IN HUMID REGIONS

Rocks which are exposed to the atmosphere are unstable. They weather chemically and mechanically into ever smaller particles which are moved downslope by **mass wasting (slides, slump, creep, and so on)**, and then are readily picked up by streams. Landform development in humid regions, where water is abundant, is characterized by the progressive deepening and widening of river valleys by stream processes and the rounding of hills by mass wasting. Processes of landform development are continuous but the discussions below are arbitrarily divided into stages. The model is an ideal, assuming a region of tectonically stable, horizontal strata.

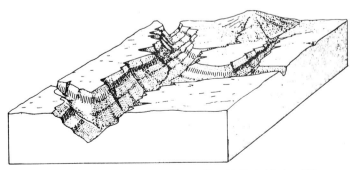

Reprinted by permission from Paul E. Myers, *Laboratory Manual in Physical Geology,* 1966

YOUTHFUL STAGE

Beginning with a roughly horizontal land surface, weathered at the surface, running water begins the process of landform development by picking up and removing the weathered surface materials. Rainwater initially flows off the surface in sheets, but soon accumulates in rivulets and streams. The major erosive process is down cutting (**vertical erosion**) Valleys have a tendency to be steep sided ("V"–shaped) and fairly straight. The initial length of these streams is short, but there is a more or less continuous tendency for each stream to increase in length by **headward erosion** (that is, the point of origin of each stream—the head—migrates in an upstream direction). This represents the earliest stage in topographic development in humid regions.

MATURE STAGE

As stream valleys form, more land is exposed to weathering along the valley walls. The walls sag under the pull of gravity and slump and collapse into the stream where water carries the debris away. As a result the initial steep valley walls become more and more rounded. As more tributary streams form and grow by headward erosion, more of the countryside is exposed to mass wasting and rounding. In time none of the initial flat upland will remain. Down-

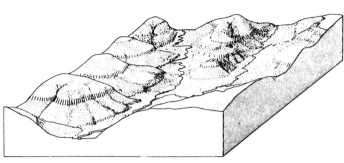

Reprinted by permission from Paul E. Myers, *Laboratory Manual in Physical Geology,* 1966

stream smaller streams become tributaries to the larger, initial streams. A pattern is developing. Smaller streams flow into larger streams, and these in turn flow into even larger streams. Each larger stream has about an equal number of tributary, or secondary streams, and so forth. This is referred to as **integrated drainage** and from the air, or on a topographic map, it creates a distinctive **dendritic** (treelike branching) stream pattern which can be readily recognized in both the youthful and mature stages of development.

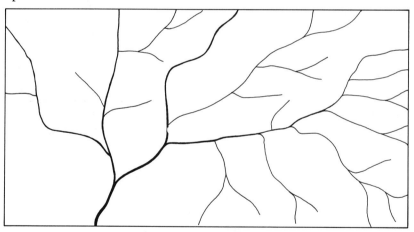

DENDRITIC DRAINAGE PATTERN

At this mature stage the larger streams begin to cut sideways and meander in a sinuous pattern. The meandering streams undercut the valley sides leading to increased slumping and other mass wasting, which tends to widen the valley. During floods, as the streams overflow their banks, sediment is deposited on the valley floor, helping to widen it, but also making it flat bottomed. This stage, with rounded hills and meandering rivers, is generally referred to as the **mature stage.**

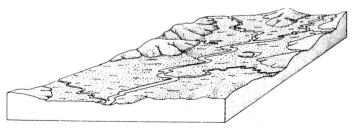

Reprinted by permission from Paul E. Myers, *Laboratory Manual in Physical Geology,* 1966

OLD AGE STAGE

The processes operating during the development of maturity continue. The hills progressively diminish in size, and the rivers meander over wider and wider regions. Steady erosion and removal of sediment lowers the land closer and closer to sea level, or *base level*, the lowest point erosion can go to. Eventually, hills are few or non-existent. Those low rounded hills that do remain are called **monadnocks,** after Mt. Monadnock in New Hampshire. At this stage the rivers are no longer just eroding. There is more sediment available than they can carry. As a result, they deposit sediment (called alluvium) on the inside of the bends of rivers and erode on the outside of the bends. The sediment is, thus, alternately deposited and eroded as the river meanders across the land, moving downstream toward base level in fits and starts. When nearly all the land is near base level, a **peneplain** has been formed. This land of large meandering rivers, a well-developed peneplain, poor drainage, and occasional monadnocks is **old age.**

SUMMARY OF LANDFORMS IN HUMID REGIONS

FEATURE	YOUTH	MATURITY	OLD AGE
CHANNEL	Nearly straight	Some meanders	Widely meandering
VALLEY WALLS	Steep	Moderate	Gentle and low
CROSS SECTION			
DEPOSITION	Little	Narrow Flood Plain	Much deposition; broad flood plain
EROSION	Downcutting	Lateral cutting begins	Lateral cutting in alluvium
STREAMS	Few—straight; "V" shaped	Many—meandering in rounded valleys	Few—meandering across wide plains
DRAINAGE	Poorly developed	Excellent	Poor and swampy
OTHER FEATURES	Falls, rapids	Sometimes rapids; small meanders	Swamps, oxbow lakes

INDIVIDUAL STAGES IN THE DEVELOPMENT OF HUMID LANDFORMS

☐   (1)   Work together in groups of two or three.

☐   (2)   Use maps provided by your instructor, or obtain the following maps from the back of the manual: "Leavenworth Quadrangle, Missouri-Kansas";   "Philipp Quadrangle, Mississippi"; "St. Paul Quadrangle, Arkansas"; and "Jacksonville Quadrangle, Illinois."

☐   (3)   Arrange the maps on your desk, left to right, according to what you believe to be stages of development from *youth* to *maturity* to *old age*. Since the stages of development between youth, maturity, and old age are gradational, it is not uncommon to designate early, middle, and late for each stage (*for example,* early, middle, and late maturity). Two of the regions you will examine represent different degrees of development of one stage;   arrange them in order of development for that stage.

☐   (4)   When you have what you believe to be the right order, write the names of the maps for each area in order below:

☐   (5)   Finally, you have just taken maps of the four regions above and arranged them in order from youth to old age. To do so you made some interpretations of the topography and judgments of their meaning. Many such judgments are made intuitively—we do it unconsciously. But it is important not only to understand intuitively, but also to recognize specific pieces of evidence by which the interpretations are made. We would like you to go back now to each individual area and pinpoint some of these specific pieces of evidence, both to clarify your own intuition and for future reference.

    A.   Begin with the most youthful region and work, map by map, to old age.
    B.   For each region pinpoint several (two or three) pieces of evidence in the form of specific topographic features which are definitive for or indicative of its particular stage of development.
    C.   In the boxes provided on pages 247 and 248, concisely and definitively, describe each topographic feature (sketch a profile), and indicate its location by longitude and latitude, township and range, or some other convenient means (e.g., northwest corner of the map) so that you could quickly locate it again.
    D.   Ask your instructor to check your arrangement and evidence for that arrangement.

MAP
NAME _____ CLIMATE _____

STAGE OF
DEVELOPMENT _____

DESCRIPTION OF TOPOGRAPHIC FEATURE	LOCATION OF TOPOGRAPHIC FEATURE
1.	1.
2.	2.
3.	3.

MAP
NAME _____ CLIMATE _____

STAGE OF
DEVELOPMENT _____

DESCRIPTION OF TOPOGRAPHIC FEATURE	LOCATION OF TOPOGRAPHIC FEATURE
1.	1.
2.	2.
3.	3.

MAP
NAME _____ CLIMATE _____ STAGE OF
DEVELOPMENT _____

DESCRIPTION OF TOPOGRAPHIC FEATURE	LOCATION OF TOPOGRAPHIC FEATURE
1.	1.
2.	2.
3.	3.

MAP
NAME _____ CLIMATE _____ STAGE OF
DEVELOPMENT _____

DESCRIPTION OF TOPOGRAPHIC FEATURE	LOCATION OF TOPOGRAPHIC FEATURE
1.	1.
2.	2.
3.	3.

Part Two

LANDFORM DEVELOPMENT IN SEMIARID BLOCK FAULTED REGIONS

Unlike the model of humid landform development where we assumed abundant rainfall and a tectonically stable situation, this model of landform development assumes both a semiarid climate and tectonic activity in the form of horst and graben faulting along normal faults. Such landforms are typical of places in Nevada, Utah, and adjoining states in the southwest.

A major difference in landform development between humid and semiarid climates is, of course, the general lack of water most of the year in semiarid climates. As a result, different processes of weathering and erosion dominate in semiarid regions. Chemical weathering, for example, is lower and coarse sands and gravels dominate the sediment. Rainfall, when it does come, is often torrential and short-lived, producing **flash floods** after which the water quickly evaporates or sinks into the ground and disappears. In a flash flood the loose sand and gravel are picked up, carried, and deposited rapidly and helter-skelter in **ephemeral streams** (streams which are dry most of the time and flow only during rains in the immediate area) or spread out as **sheet flow** (thin sheets of water and sediment which flow across the surface rather than in a channel). Sometimes the sediment and water thicken into heavy **mudflows** that flow like concrete sliding down the chute of a concrete truck, "freezing" in mounds and blobs along the slope.

Although the rains in a semiarid region may be intermittent or seasonal, over long periods of time considerable amounts of sediment can be moved by these floods. Instead of being carried to the sea as in humid regions, however, the sediment accumulates in large deposits near the source. One difference, then, between humid and semiarid regions is that humid regions tend overall to be environments of sediment removal and lowering of topographic elevation whereas semiarid regions tend to be environments of sediment accumulation. The model below describes the progressive development of landforms in regions of tectonic activity and semiarid climates.

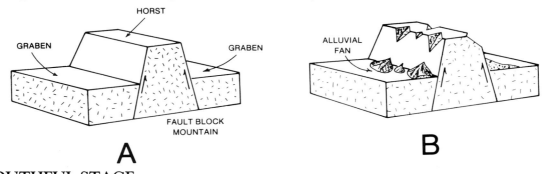

YOUTHFUL STAGE

Imagine a mountain formed by a block of crust (a horst) faulted above the surrounding valleys (graben)—a **fault block mountain** (see A above). Material eroded from the top of the horst is carried over the edge of the fault block and is dumped near the base because the water carrying it rapidly spreads out and infiltrates. The sediment accumulates in radiating, wedge-shaped features called **alluvial fans** (see B). In addition, because of the lack of chemical weathering, the topography remains sharp and angular. Compare this with the rounded hills and valleys in humid regions.

SUMMARY OF LANDFORMS IN SEMIARID BLOCK-FAULTED REGIONS

FEATURE	YOUTH	MATURITY	OLD AGE
TOPOGRAPHY	Rugged topography; few smooth sloping plains	Development of smooth, sloping plains	Mostly smooth slopes, isolated inselbergs
ALLUVIAL FANS	Small and individual	Beginning to coalesce	Fully intergrown
INSELBERGS	Not present	Not present	Present
FAULTS	Visible as a linear cliff or scarp	Fault scarp discontinuous; partially buried	No fault scarps visible
PLAYAS	May be present	May be present	May be present
STREAM CHANNEL	Straight	Straight	Straight

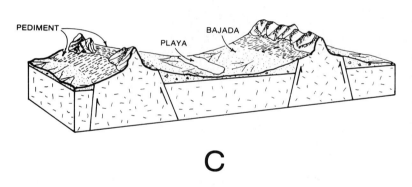

C

MATURE STAGE

Erosion of the horst block continues. Numerous alluvial fans form and grow in size, eventually coalescing (growing together), losing their individuality, and building broad-sloping sedimentary deposits piled up parallel to the base of the horsts; these are **bajadas** (see C). You will notice that as the sediment pile grows, the horst is slowly being buried by its own debris. This is the mature stage.

OLD AGE STAGE

Continued erosion and deposition bury the horst more and more until only small isolated peaks (**inselbergs**) surrounded by broad, gently sloping plains

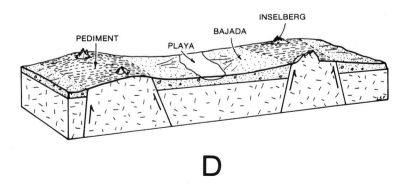

D

fanning out in all directions remain. In the final step even the inselbergs will disappear completely, and what remains is a broadly domed, flat or gently sloping surface called a **pediment** (see D). Although the bajada and pediment are almost indistinguishable on the surface, the pediment is underlain by solid bedrock and the bajada is underlain by loose sediment. This is the **old age stage.**

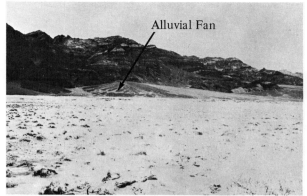

Youthful Stage

Mature Stage

Old Age Stage

Three stages in the development of semiarid landforms associated with fault block mountains.
Photos by Eliot Blackwelder.

The formation of a fault block mountain is not usually an isolated occurrence; in fact, it is more usual to find swarms of alternating horsts and adjacent graben forming together, as in the southwestern United States. In these circumstances landforms characteristically associated with isolated horsts do develop, but in the presence of more than one horst the process of erosion and sediment accumulation results in the development of an additional new feature. Because of the intermittent and infrequent rainfall, a drainage system of permanent streams and rivers does not develop. Sediment accumulating first near the mountains and then spreading out flood by flood begins to fill in the basins between the horsts. Stream channels which do develop become choked with debris, as the water from each rainstorm sinks into the ground, dumping its sediment load. In this way, even those low areas at the end of each basin (where drainage to the sea would normally develop) begin to build up. The result, in time, is a completely enclosed basin blocked on the sides by the mountains and on each end by accumulated sediment. During especially heavy rains water from the surrounding mountains accumulates in these basins, forming **playa lakes** (see D), the water stagnates and evaporates, and dissolved salts brought from the surrounding land increase in concentration (e.g., the Great Salt Lake, Utah) and precipitate out as salt deposits. This process of playa formation is unique to semiarid regions and may occur at any stage of development. In association with alluvial fans and bajadas, it is a diagnostic climatic indicator.

INDIVIDUAL STAGES IN THE DEVELOPMENT OF SEMIARID LANDFORMS

☐ (1) Work together in groups of two or three.

☐ (2) Use maps provided by your instructor, or obtain the following maps from the back of the manual: "Antelope Peak Quadrangle, Arizona"; "Ennis Quadrangle, Montana"; "Furnace Creek Quadrangle, California"; and Polvadero Gap Quadrangle, California."

☐ (3) Arrange the maps on your desk, left to right, according to what you believe to be stages of development from youth, to maturity, to old age. Since the stages of development between youth, maturity, and old age are gradational. it is not uncommon to designate early, middle and late for each stage (for example, early maturity, middle maturity, and late maturity). Two of the regions you will examine represent different degrees of development of one stage; arrange them in order of development of that stage.

☐ (4) When you have what you believe to be the right order, write the names of the maps for each area in order below:

SEMIARID >

YOUTH

↓

OLD AGE

1. _____

2. _____

3. _____

4. _____

☐ (5) Finally, as in the analysis of landform development in humid regions, your interpretation of the sequence of landform development in semiarid regions requires an understanding of the theory of the processes involved, and the ability to interpret their topographic expression. We would like you to again pinpoint some specific evidence for your arrangement on pages 253 and 254.

A. Following the same directions as in 5B, C, and D on page 246 for the specific analysis of humid landform features, support your arrangement of the semiarid landform regions.

MAP
NAME _____ CLIMATE _____

STAGE OF
DEVELOPMENT _____

DESCRIPTION OF TOPOGRAPHIC FEATURE	LOCATION OF TOPOGRAPHIC FEATURE
1.	1.
2.	2.
3.	3.

MAP
NAME _____ CLIMATE _____

STAGE OF
DEVELOPMENT _____

DESCRIPTION OF TOPOGRAPHIC FEATURE	LOCATION OF TOPOGRAPHIC FEATURE
1.	1.
2.	2.
3.	3.

MAP
NAME _____ CLIMATE _____ STAGE OF
DEVELOPMENT _____

DESCRIPTION OF TOPOGRAPHIC FEATURE	LOCATION OF TOPOGRAPHIC FEATURE
1.	1.
2.	2.
3.	3.

MAP
NAME _____ CLIMATE _____ STAGE OF
DEVELOPMENT _____

DESCRIPTION OF TOPOGRAPHIC FEATURE	LOCATION OF TOPOGRAPHIC FEATURE
1.	1.
2.	2.
3.	3.

Part Three

SEMIARID LANDFORMS—VARIATIONS ON THE CLASSIC PATTERN

The study of semiarid landforms began in the North American west, where fault block (horst) mountains commonly occur. With their distinctive alluvial fans, bajadas, and inselbergs, horst mountains acquired special importance in the theories of landform development. But horst mountains are certainly not the only circumstances in which arid landforms develop. The underlying geology is basic to determining exactly which landforms develop. Four distinct situations can be defined.

1. **Horst mountains** with developing alluvial fans, bajadas, pediments, inselbergs, and playas.
2. **Badlands** developing in areas underlain by fine-grained, easily erodable horizontal strata, all of equal resistance to erosion.
3. **Mesas** and **buttes** developing in areas underlain by horizontal strata but capped by a layer of resistant rock.
4. **Hogbacks** and **cuestas**[1] developing in areas underlain by dipping strata, usually with rock layers of varying resistance to erosion.

Landform development in both humid and semiarid climates results from fundamentally the same processes—stream erosion and mass wasting. The difference, of course, is one degree, summarized in the chart below.

HUMID CLIMATES	SEMIARID CLIMATES
1. Frequent rains with continuously flowing streams	1. Long periods with no rain (streams dry) followed by torrential rains
2. Good vegetation cover to hold the soil	2. Sparse vegetation; soil easily eroded by heavy rain
3. Chemical weathering extensive; accompanied by active mass wasting	3. Chemical weathering very slow; mass wasting not as important as stream erosion
4. Erosion predominates in creation of landforms	4. Depositional landforms often form major part of topography in valleys

[1] Strictly speaking, landforms with hogbacks and cuestas may develop in humid climates, also, since their presence is more strongly controlled by underlying rock structure than by climate.

THE DEVELOPMENT OF BADLANDS

Badlands form in regions underlain by fine-grained, easily erodible, horizontal strata all of equal resistance to erosion. They are characterized by extremely rough, high, steeply gullied topography.

South Dakota Badlands. Photo by L.S. Fichter.

In the simplest case, badlands are a boundary or transition zone between a plateau area and a drainage (stream) system. It is the zone of most active erosion, occurring principally during torrential rains.

The illustrations on page 257 are a series of cross sections showing the progressive development of a typical badland area. The first cross section (A) shows the initial stage. A broad flat plateau exists but is cut by a youthful stream valley, dry most of the year (that is, an ephemeral stream). The plateau is probably covered by grass or low scrub plants which protect the land, to some degree, during the torrential rains. The banks bordering the stream, however, are generally not vegetated, since the torrential rains strip the slopes of their plant cover and remove any soil that is present. It is at this point that the badlands begin to develop.

Cross-section B shows the beginning of badland development. Tributary streams form during torrential rains, and rapid vertical and headward erosion occurs. With each rain, these valleys are extended by vertical and headward erosion, and the area is further dissected as in cross-section C. These tributaries are in the form of numerous small streams which finely dissect the landscape. In cross-section C, the main stream forms the base level for the developing badlands. The tributary streams cannot erode below the level of the main stream, and their disgorged sediment, stripped in large quantities from the bare slopes, is spread over the valley of the main stream in a sheet-like deposit (a bajada).

Cross-section D shows a further development of these processes. A typical area with badland topography, thus, has a flat plateau bordered by the transitional badlands. Between the badlands and the main stream is the gently sloping area underlain by sediment separated by a sharp break between the bedrock and bajada.

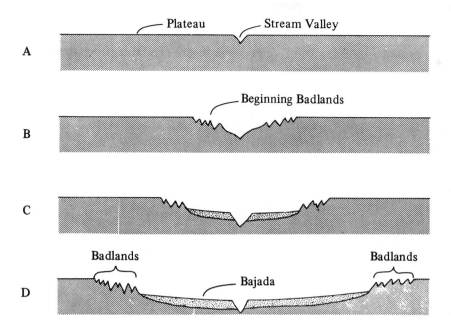

DEVELOPMENT OF MESAS AND BUTTES

Mesas and **buttes** develop in semiarid regions by the erosion of horizontal strata capped by a layer (or layers) of resistant rock.

Mesas are broad, flat-topped hills or mountains bounded by steep cliffs with accumulations of loose, weathered debris at their base (**talus**). Buttes are isolated remnants of dissected mesas, which may or may not have flat tops, depending on how much erosion has reduced their size. Nonetheless, they typically still retain their steep sides and accumulated talus at their base.

Mesa and butte (isolated remnant near center of picture) near Mesa Verde, Colorado. Photo by L.S. Fichter.

The distinction between badland development and the formation of mesas and buttes is, of course, the absence (badlands) or presence (mesas/buttes) of resistant caprock. Nonetheless, the development of mesas and buttes begins with the stream dissection of a plateau.

Cross section A below shows the first stage in the development of mesas and buttes. Small streams have begun to cut into the resistant caprock beginning at the edge of the plateau and eroding across the plateau by headward erosion. As downcutting progresses, the streams cut narrow, steep-sided gorges; "V"-shaped valleys do not form because chemical weathering is not as prominent in semiarid region and the caprock is not significantly affected by mass wasting; that is, hillside rounding, typical of humid regions, does not occur. The steep sides are maintained by the resistant caprock as the canyon becomes wider. Canyon widening occurs when the softer rock beneath the caprock erodes away, undercutting and weakening it. Joints eventually develop in the caprock until the edge breaks loose and falls to the base, accumulating as talus (see cross sections B and C). As in other areas of semiarid landform development, sediment accumulates in the valleys between buttes and mesas. Sediment accumulation forms nearly flat valley floors, gently sloping toward the main stream channel; except during (and right after) rains, these stream channels are dry gullies. Cross section D shows a fully developed mesa and butte.

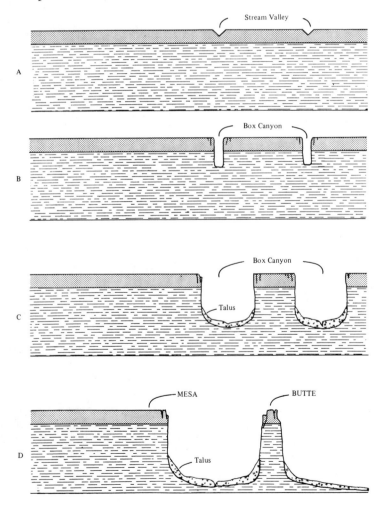

DEVELOPMENT OF HOGBACKS AND CUESTAS

Hogbacks and **cuestas** develop in areas underlain by dipping strata, usually with rock layers of varying resistance to erosion.

Hogbacks are long, narrow ridges with steep slopes on both sides. Cuestas are hills or ridges with a long, gentle slope on one side, and a short, steep slope on the other side.

Grand Hogback, Colorado. Photo by D. Poché.

Cuesta southwest of the Nacimiento Mountains, New Mexico. By permission of Ward's Natural Science Establishment.

The distinction between hogbacks and cuestas is one of degree, and it is quite possible for these two landforms to grade into each other. The controlling factor is the amount of dip of the rock layers. The rocks forming hogbacks generally dip at a steep angle (roughly 30°–90°), while those forming cuestas dip at a lesser angle.

As with badlands, mesas, and buttes, hogbacks and cuestas develop as a result of stream erosion. And, as in the case of mesas and buttes, the resistant rock layers strongly influence the developing landforms. These rock layers weather and erode more slowly, "capping" the ridge and forming the slope parallel to the dip. The illustrations below show these relationships.

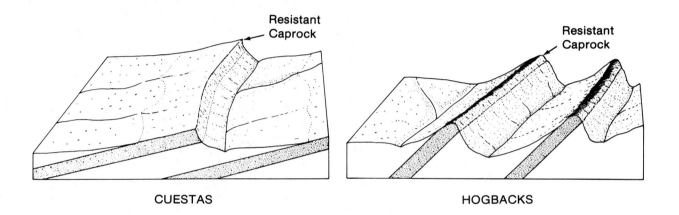

CUESTAS   HOGBACKS

VARIATIONS ON SEMIARID LANDFORM DEVELOPMENT

☐ (1) Work together in groups of two or three.

☐ (2) Use maps provided by your instructor, or obtain the following maps from the back of the manual: "Egloffstein Butte Quadrangle, Arizona"; "Loveland Quadrangle, Colorado"; "Sheep Mountain Table Quadrangle, South Dakota"; and "Wade Quadrangle, Montana."

☐ (3) Examine each of the areas represented by the *four* maps and determine whether they illustrate badlands, mesas, buttes, hogbacks, or cuestas. More than one of these landforms may be found on one map.

☐ (4) Finally, as in previous analyses of landform development, pinpoint some specific evidences for your decisions and record them and their locations in the chart on page 261.

☐ (5) Ask your instructor to check your conclusions and evidence for those conclusions.

MAP NAME	LANDFORM	EVIDENCE AND LOCATION
_____		
_____		
_____		
_____		

┌─ *Part Four* ──────────────────────────────────┐
│   │
│ _____   **REJUVENATION** _____   │
└───┘

   The patterns of landform development for humid and semiarid regions were described in the ideal, theoretical state. As you have seen from your examination of various parts of the country, the ideal pattern is not necessarily obvious. The numerous variations are the result of many different causes, such as bedrock type, unusual structural development, or fluctuating climatic conditions.

   In areas you have not yet observed, unusual and restricted erosive effects like glaciers, or special bedrock types like the presence of easily dissolved limestone, may produce dramatic differences; we want to examine the effects of some of these variables. For the moment, we want to look at the special case of interruption in landform development occurring in humid, and certain semiarid, regions.

   You will recall that in the final stages of humid landform development, the land surface is reduced to near base level—the **peneplain.** The bedrock is covered with a thin veneer of sediment which is worked and reworked by the now sluggish and broadly meandering rivers (for example, the old age stage of humid development). Looking at this, you will not see any special, outstanding features, but below the sediment veneer there may exist a complex of structurally deformed igneous, sedimentary, and metamorphic rocks (for example, the roots of an old mountain range).

Throughout the history of the earth, it has been unusual for any part of the earth to remain tectonically stable for any extensive period of time. Peneplained regions are, thus, often lifted above base level again. As this uplift occurs, the previously sluggish rivers increase their ability to erode. They quickly cut through and remove the veneer of sediment from their channels and begin cutting solid bedrock. Normally, when a river or stream is forming, its channel will select the softest rock because it is most easily eroded; thus the straight, steep-sided channels typical of the *youthful stage* form first. But during uplift of the peneplain, the broadly meandering rivers cut down vertically in their preestablished meandering courses regardless of the varying resistance of the underlying bedrock. The result is an old age river (broadly meandering) in a youthful channel (steep-sided or "V"-shaped). Such rivers which have been trapped in their channel by uplift are referred to as **entrenched.** But normal surface processes, such as weathering, slumping, and landsliding, begin to occur again also. The harder, more resistant rocks, because they are less easily weathered and eroded, form hills or ridges while the softer rock in between is removed to form valleys. The original river does not follow this new pattern of landform development but cuts across all of these ridges and valleys equally well. The result is the formation of a **water gap**—a deep ravine across a mountain ridge that allows the passage of a stream *through* the mountain rather than around it, a sure indicator of uplift from a peneplain.

And just as diagnostic, in areas where folded rocks have been peneplained and uplifted, a distinctive drainage system develops to accompany the water gaps. It is called a **trellis** pattern, characterized by parallel running major streams separated by mountain ridges, with their approximately right-angle tributaries forming a latticework. (See illustration below.)

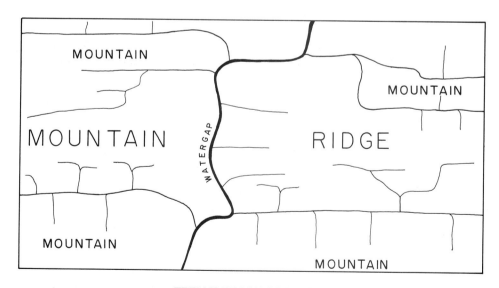

TRELLIS DRAINAGE PATTERN

In areas of undeformed sedimentary rocks, uplift produces a **plateau** (a large, nearly flat mountain of horizontal strata). Old rivers flowing across a rising plateau rapidly entrench themselves, forming a deep meandering scar across the land. But because the surface of the horizontal strata is everywhere equally resistant to erosion, the tributaries to the entrenched river, which now begin to dissect the plateau, develop a dendritic drainage pattern (see illustration, page 244).

Regardless of the cause, the uplift of an old land area accompanied by river entrenchment and differential erosion creates new distinctive landforms. The sum total of all these processes is referred to as **rejuvenation.**

CASES OF REJUVENATION

☐   (1)   Work together in groups of two or three.

☐   (2)   Use maps provided by your instructor, or obtain the following maps from the back of the manual: "Anderson Mesa Quadrangle, Colorado"; "Harrisburg Quadrangle, Pennsylvania"; and "Mammoth Cave Quadrangle, Kentucky."

☐   (3)   The territory covered by each of these maps represents a case of rejuvenation under different geological circumstances.
       A.   One region is formed by rejuvenation of a plateau with a strongly entrenched meandering river and beginning dendritic drainage pattern in the tributaries.
       B.   A second is formed by rejuvenation of an area underlain by the roots of an old thrust-faulted and folded mountain range showing strong structural deformation, with rocks of varying resistance to weathering and erosion, water gaps, and a trellis drainage pattern.
       C.   A third is formed by rejuvenation of areas underlain mainly by limestone in which many rivers flow underground in subterranean river channels for some of their length (thus, often mysteriously disappearing from the surface), and where roughly circular **sinkholes** form at the surface by the collapse of cave ceilings.

☐   (4)   Examine each of the areas represented by the three maps and determine whether they represent the plateau, deformed rock, or area underlain by limestone; list their names in the table below.

REJUVENATION	PLATEAU —
	DEFORMED ROCKS —
	UNDERLAIN BY LIMESTONE —

☐   (5)   In the boxes provided on the following page, briefly and concisely describe how you identified each territory; use such evidence as general overall topography, specific topographic features (for example, sinkholes, water gaps), profiles, or any other kinds of supporting evidence. Make sketches of the topographic features on the map which identify each geologic condition.

REJUVENATED RIVER _____   MAP NAME _____

EVIDENCE THAT ROCKS ARE HORIZONTAL

PLATEAU

REJUVENATED RIVER _____   MAP NAME _____

EVIDENCE THAT ROCKS ARE DEFORMED

DEFORMED ROCKS

REJUVENATED RIVER _____   MAP NAME _____

EVIDENCE OF LIMESTONE EROSION

UNDERLAIN BY LIMESTONE

Glaciation–
Accentuation and Obliteration

PURPOSE

Topographic maps, as we see them, represent a single instant (in terms of geologic time) of the interaction between the rocks of the earth and the atmosphere. It would be much more interesting if we could watch landforms develop through many instances of geologic time in the form of a motion picture; we could then see that landform development results from a sequence of highly integrated, continuous, and somewhat more believable processes. Normally we are not able to observe the forces which sculpt the land actually operating in this highly integrated way, primarily because they are extremely slow, short-lived, and/or difficult to observe. Glaciers are the exception; although they may be extremely slow, they carry the results of their activity with them, and are around for more than enough time to be thoroughly studied. Today we observe glaciers at work and try to make a connection between the glaciers themselves and the special landforms they create.

There are two parts to this laboratory:

PART ONE—ALPINE GLACIATION

PART TWO—CONTINENTAL GLACIATION

┌─ Part One ─────────────

ALPINE
GLACIATION

INTRODUCTION

The normal development of landforms results from the interaction of water and the earth's surface. The different landforms that you have seen developing have resulted from variations in the amount of rainfall, the degree of chemical weathering and mass wasting, and tectonics. Throughout much of earth's history, these processes have been the dominant forces of landform development.

There have been, however, intervals in earth history appearing about every 250 million years when extraordinary processes have influenced the earth—the worldwide lowering of the earth's average temperature and the accumulation of vast sheets of ice (glaciers). The mechanism(s) responsible for the development of glaciers is not yet clearly understood, but the result is the accumulation of enormous sheets of ice in the polar and mountainous regions of the earth. And in time these sheets of ice grew and moved thousands of miles across the continental masses into mid-latitude regions.

Ice is normally thought of as a brittle substance, but in huge masses it flows and expands outward from the area of greatest accumulation. Currently, the ice masses of the world are diminishing, and land which was once covered by glaciers is now exposed. In areas where these glaciers were once active, however, highly specialized and unique landforms developed. These landforms have resulted from a complex and unique combination of erosional and depositional processes which are not separated from each other in any systematic way. Erosional and depositional features may be jumbled together in an area, unlike the features formed by running water. These areas have in the past and still today cover substantial portions of the earth and have economic, cultural, and scenic significance.

Ice, because it is a solid, can carry completely unsorted debris of all sizes from huge boulders to clay, a process very unlike river erosion and transportation. As the glacier moves, it bulldozes rock material in front of it. In mountainous regions rock material is added to the ice by mass wasting (mechanical weathering, such as **frost wedging**). Rock fragments falling on top of the glacier will eventually sink to the bottom. As the ice moves, the rock debris at its base and at its front (**snout**) acts as a powerful abrasive, scouring the land and creating landforms which could not develop in any other way. When the ice retreats, material is dumped helter-skelter in piles or sheets.

Unglaciated (note "V"-shaped valleys).

Glaciers, fully developed.

Glaciated valley with the ice melted (note "U"-shape).

Glaciation occurs in mountainous and nonmountainous regions. It is characteristically separated on this basis into **mountain** (or **alpine**) and **continental glaciation**, although these may intergrade in certain areas. Some resulting landforms are typical of both regions, while others are unique to one or the other. We'll look at examples of both.

☐   (1)   Work together in groups of two or three.

☐   (2)   Use maps provided by your instructor, or obtain the following maps from the back of the manual: "Chief Mountain Quadrangle, Montana"; and "Cordova Quadrangle, Alaska."

☐   (3)   Both the Chief Mountain and Cordova quadrangle maps represent mountain or alpine glaciation. In the Cordova area, the glaciers are still present, while in the Chief Mountain area only a few small glaciers remain. In humid areas where glaciers have not been active, processes of weathering, erosion, and deposition occur very slowly and are often undramatic. In contrast, by examining Cordova and Chief Mountain areas we can see glaciation in action and the results of glaciation. First, we will examine the processes of glaciation in the Cordova area.

CORDOVA QUADRANGLE

Much of this area is covered by snow and ice. The presence of snow and ice in itself, however, is insufficient to produce dramatic landforms. Only ice which has piled up enough to begin moving is capable of bulldozing and scouring the landscape. It is often difficult to determine which bodies of ice are moving, but presumably the largest bodies found in this area are capable of moving and, in fact, are moving down valley. But you don't have to take our word for it; the ice provides its own evidence. There are at least three pieces of evidence on the map that show that the large glaciers and their tributaries (for example, Scott, Sheridan, and Sherman glaciers) are, in fact, moving.

☐   (1)   Ice, like streams, moves down valleys; think about the process of flow. Where is the flow most rapid, where is it slowest? Examine the contour lines (in blue) running across the ice. (Do they differ from the head to the snout of the glacier? How are they similar to or different from contour lines in an unglaciated valley? Is the glacier carrying sediment? What is the evidence? How is the sediment distributed? How did it get that way?) In the spaces provided on page 268, describe the pieces of evidence for flow that you have found.

☐ (2) When you have finished, ask your instructor to check your evidence.

EVIDENCE OF ICE MOVEMENT ON THE CORDOVA QUADRANGLE

DESCRIPTION OF EVIDENCE	LOCATION
1.	
2.	
3.	

In high mountain regions where snow is abundant, glaciers begin as **snow fields** in **cirques.** If the summer melting season does not cause all the snow to melt, isolated snow fields remain. Small, at first, they grow larger year by year. The first snow accumulations are compressed through several stages to form **ice fields.** Erosion and weathering (through frost wedging) in these fields begins the process of glaciation. The small basins developing in these areas are cirques. All mountain glaciers begin in this way.

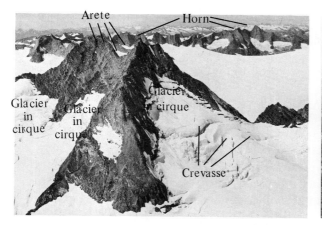 

Horns, arêtes, and cirques. Photo on left shows a mountain in the process of being sculpted by glaciers. Photo on right shows another glaciated mountain but with the ice melted. U.S. Forest Service.

☐ (1) On the Cordova quadrangle, find several cirques. In some of them the ice may be moving; in others it may not be moving. Can you determine which are and which are not moving?

CIRQUES ON THE CORDOVA QUADRANGLE

EVIDENCE OF ICE MOVEMENT	LOCATION
1.	
2.	
3.	

Cirques will begin developing in almost any hollow. They frequently are found in bunches on mountain flanks. When a series of cirques filled with ice and resulting glaciers surround a mountain peak, they will often sculpt a steep, sharp-pointed, angular **horn** (for example, the Matterhorn in Switzerland). As the cirques and glaciers increase in size, sharp angular **arêtes** (narrow, rocky, sharp-edged ridges which are sculpted by glaciers and which separate cirques) develop.

☐ (1) Locate by any means available (longitude/latitude, township/range, etc.) several horns and/or arêtes and describe them in the space provided.

HORNS AND ARÊTES ON THE CORDOVA QUADRANGLE

HORN OR ARÊTE	LOCATION
1.	
2.	
3.	

Stream-developed valleys in mountainous regions are normally "V"-shaped, characteristic of youth. Glaciation obliterates this kind of valley development.

☐   (1)   Sketch a freehand profile across one of the large glaciated valleys (down-valley from a major glacier).

PROFILE OF A VALLEY ON THE CORDOVA QUADRANGLE

VALLEY NAME

CHIEF MOUNTAIN QUADRANGLE

From your examination of the Cordova area, you should intuitively recognize this as an area of alpine glaciation.

☐   (1)   Locate by any means available (longitude/latitude, township/range, etc.) an example of each of the features of alpine glaciation listed in the chart below.

ALPINE GLACIAL FEATURES ON THE CHIEF MOUNTAIN QUADRANGLE

FEATURE	LOCATION
CIRQUE WITHOUT ICE	
HORN	
ARÊTE	
"U"-SHAPED MOUNTAIN VALLEY	

GLACIAL DEPOSITIONAL FEATURES, CHIEF MOUNTAIN QUADRANGLE

MORAINES AND GLACIAL OUTWASH   Compared to landforms typical of humid or semiarid regions, the sharply sculpted mountains and broadly scoured valleys in the Chief Mountain area indicate a large amount of erosion has occurred here. Most glacial weathering is mechanical, resulting from frost wedging and glacial scouring and plucking (when the ice flows around a rock projection and tears it out), and produces an immense volume of sediment all classified as **drift.** Much of the sediment is moved by the ice itself, either at the base, within, or riding piggyback on top. These sediments are a completely unsorted and unstratified, disorganized jumble of all sizes and shapes of sedimentary particles from clay-size to boulders the size of cars or houses, all referred to as **till.** The disorganization is because ice carries all particles indiscriminantly and has no mechanism to sort the sediment as it is transported. Distinctive features of the larger particles, however, are **glacial striations**, grooves in the rocks resulting when they scrape each other during transport. Eventually this sediment gets deposited, usually during a phase when the glacier is melting backward faster than it is moving forward. These ice-laid, unsorted, and unstratified deposits are **moraines.** Several kinds of moraines are defined, depending on their location in the glacier.

> **Terminal moraine**—irregular linear piles of sediment deposited at the very front of the farthest advance of a glacier.
>
> **Ground moraine**—sediment deposited as a sheet underneath a glacier, usually after the glacier has stopped moving and is melting; seen as irregular but generally horizontal blanket deposits.
>
> **Recessional moraine**—irregular linear piles of sediment deposited behind the terminal moraine during a time when the glaciers' forward movement and backward melting are in equilibrium.
>
> **Lateral moraine**—irregular linear piles of sediment deposited at the sides of a glacier, especially an alpine glacier.
>
> **Medial moraine**—irregular linear ridge(s) of sediment formed where separate glaciers have joined and flowed together.

In addition to the unsorted and unstratified morainal sediment, **glacial outwash** is also produced when meltwater picks up and transports the morainal material. These deposits are usually braided river deposits near the glacier and are indistinguishable from river deposits produced in non-glaciated regions. Glacial outwash is recognized because of its association with other glacial features (see the photographs on page 272).

GLACIAL LAKES   Lakes commonly occur in glaciated valleys. Thinking about how glacial sediment is transported and deposited, how do you think these lakes form? Sometimes a series, or chain, of small **paternoster lakes** can be seen distributed along a glaciated valley.

☐   (1)   Examine both the Cordova and Chief Mountain areas again. Find examples of moraines and glacial outwash (there will be no map symbols showing their presence). Write their location and your evidence on the charts on pages 273-274.

Pleistocene glacial outwash in a gravel pit west of Little Falls, New York. Photo by G. T. Farmer.

Glacial outwash on the Unuk River, Canada. U.S. Forest Service.

Alpine glaciated regions thus have a great variety of erosional and depositional features, usually in close association. The arrangement is not random, however; certain features occur in certain places because of the processes which produce them. If you needed to find any particular feature (the glacial outwash may have concentrations of gold nuggets, for instance), a lot of time could be wasted without knowing where to look. But your understanding of the processes and responses involved would allow you to predict where these features might be found, saving much time and effort.

Part Two
CONTINENTAL GLACIATION

During several stages in the earth's 4+ billion year history, large ice sheets have covered major portions of the continents. The earth is currently in one of these stages, although most of the ice which existed only 10,000 years ago has melted. We do not know yet if the ice will return or continue to melt to nothing. During the recent glacial advance most of northern Europe, Asia, and northern North America were under an ice sheet 4 to 5 km thick. North American **continental glaciation** originated initially in the northern reaches of Canada and Alaska from several different centers. Beginning as large snow and ice fields, they grew large enough and thick enough to begin spreading under their own weight in all directions. As the ice moved across Canada, it stripped the land of its loose sediment and scoured the bedrock. Although both erosional and depositional features exist in Canada, most of the smaller-scale features in the United States are depositional. Depositional features are formed during both ice advance and retreat (ice does not actually move backward; it just ceases moving forward and melts—or forward movement and melting may be in equilibrium).

MORAINES ON THE CHIEF MOUNTAIN/ CORDOVA QUADRANGLE

MAP AND LOCATION	EVIDENCE
1.	
2.	
3.	

GLACIAL OUTWASH ON THE CHIEF MOUNTAIN/ CORDOVA QUADRANGLE

MAP AND LOCATION	EVIDENCE
1.	
2.	
3.	

Aerial view of a continental glacial moraine. Area in left half of picture is the moraine; note the hummocky topography. Area to the right is unglaciated farmland. Between Moses Coulee and Chelan, Washington. Photo by John S. Shelton.

CONTINENTAL GLACIAL FEATURES

FEATURE	MAP AND LOCATION	EVIDENCE
TERMINAL or RECESSIONAL MORAINE		
GROUND MORAINE		
DRUMLINS		
ESKERS		
KETTLES		
DERANGED DRAINAGE		
OUTWASH PLAINS		

☐   (1)   Work together in groups of two or three.

☐   (2)   Use maps provided by your instructor, or obtain the following maps from the back of the manual: "Jackson Quadrangle, Michigan"; Palmyra Quadrangle, New York"; and "Whitewater Quadrangle, Wisconsin."

☐   (3)   Read the descriptions of continental glacial landforms discussed below.

☐   (4)   Each feature listed in the chart on page 275 is found on one or more of the areas represented by the maps. Find, locate by whatever means possible (longitude/ latitude, and so on), and describe an example of each.

MORAINES

By the time the glaciers moving out of Canada had reached the United States, their snouts contained or were pushing large quantities of debris in front of the advancing ice, and additional quantities were more or less evenly scattered throughout the base. As long as ice continues to move, more and more debris will pile up and be bulldozed ahead of the ice, but as soon as the ice stagnates, its load of dirt and rock is dumped helter-skelter in linear or lobate moraines. Retreat or recession of the ice may not occur all at once but in pulses so that numerous moraine piles are formed. A variety of moraine types are described on page 271.

DRUMLINS

Drumlins are low, smoothly rounded, elongate or oval hills, usually with one end blunt (pointing in the direction from which the glacier came) and the other sloping gently. Their exact method of formation is not known, but they may be formed from a former moraine by readvancing ice.

Drumlins east of Rochester, New York. Photo by John S. Shelton.

ESKERS

With final stagnation of the ice, large quantities of meltwater are released. Rivers often flow within or under the ice. These produce unique and unusual topographic deposits. Streams which develop on land cut channels within the earth. Streams in the ice, however, deposit material in their channels above ground. As the ice disappears, these sediments remain as long sinuous (snakelike) ridges of stream deposits meandering over the surface. These are **eskers.**

Canadian esker (note sinuous ridge). Canadian Department of Mines, Geological Survey.

KETTLES

Most dirt and rock carried by the ice is unsorted an unevenly distributed; therefore, as the ice melts, it is dumped with no pattern. Often, areas of closed drainage and depressions result. Some depressions are formed by blocks of ice left buried in the moraine by the retreating glacier. As these ice blocks melt, the overlying sediment collapses, forming a **kettle**. Often these kettles fill with water, forming lakes, or through the accumulation of sediment and organic matter form **peat bogs** and eventually meadows and forests.

DERANGED DRAINAGE AND OUTWASH

Kettles are not the only result of sediment dumping by glaciers. An unpatterned and incomplete **deranged drainage** results. Streams flow in irregularly sized valleys. Deranged drainage is poor drainage; numerous swamps, bogs, and lakes develop.

Large, nearly flat **outwash plains** or valley fillings of sorted and stratified (commonly cross-stratified) sands and gravels may also be found. These are difficult to recognize on topographic maps (and you may not be able to find them) but they commonly occur where glacial meltwater reworked morainic debris.

CONCLUSION

For some time, in the early days of the birth of geology as a modern science, people did not believe that ice was capable of moving over vast distances and creating the special landforms they were studying. Yet they were left with a dilemma, for aside from the hypothesis that the features were formed during Noah's Flood, they were hard pressed to find a process capable of explaining the landform response they saw. This problem is not uncommon in any science; some phenomena are just very hard to explain when we are ignorant of the processes which produced them. In fact, this gap between what we have not yet seen happen and the results we can see is probably the single most influential stimulator of human's curiosity and activity. This presents an interesting problem, however, for there are in ancient rock records erosional features and sedimentary deposits formed during glaciations long since come and gone. If we lived in a time when no glaciers existed on the earth, and could not see them at work, would we be able to correctly interpret these rocks? This question has interesting implications when we realize what a small part of the universe we live in and how short a time we have had to observe it. In any event, could you, if you thought about it for a while, propose an alternate hypothesis capable of explaining the landforms attributed to glacial action?

Topographic Maps

Below is a list of the topographic maps that follow, and the map series they belong to. Only a portion of each map is printed. The map pages are not numbered but the maps are arranged in the following pattern:

St. David Quadrangle—first right-hand page

Humid Landforms—first four left-hand pages

Semiarid Landforms—right-hand pages two through five

Variations on Semiarid Landforms—left-hand pages five and six and right-hand pages six and seven

Rejuvenation—left-hand page seven and right-hand pages eight and nine

Alpine glaciation—left-hand page eight and right-hand page ten

Continental glaciation— left-hand page nine and right-hand pages ten and eleven

The bar graphs below are reproduced from the bottom of a 7 1/2 minute and a 15 minute quadrangle map and can be used to scale distances on the maps.

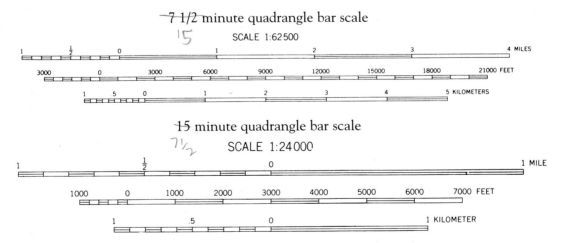

7 1/2 minute quadrangle bar scale

SCALE 1:62 500

15 minute quadrangle bar scale

SCALE 1:24 000

EXERCISE MAP FOR THE PUBLIC LAND SURVEY SYSTEM

St. David Quadrangle, Arizona–15 Minute Series

HUMID LANDFORMS

Leavenworth Quadrangle, Missouri-Kansas–15 Minute Series
Philipp Quadrangle, Mississippi–15 Minute Series
Jacksonville Quadrangle, Illinois–15 Minute Series
St. Paul Quadrangle, Arkansas–15 Minute Series

SEMIARID LANDFORMS

Antelope Peak Quadrangle, Arizona–15 Minute Series
Ennis Quadrangle, Montana–15 Minute Series
Furnace Creek Quadrangle, California–15 Minute Series
Polvadero Gap Quadrangle, California–15 Minute Series

SEMIARID LANDFORMS—VARIATIONS ON THE CLASSIC PATTERN

Egloffstein Butte Quadrangle, Arizona–15 Minute Series
Loveland Quadrangle, Colorado–15 Minute Series
Sheep Mountain Table Quadrangle, South Dakota–7.5 Minute Series
Wade Quadrangle, Montana–7.5 Minute Series

REJUVENATION

Anderson Mesa Quadrangle, Colorado–7.5 Minute Series
Harrisburg Quadrangle, Pennsylvania–15 Minute Series
Mammoth Cave Quadrangle, Kentucky–15 Minute Series

ALPINE GLACIATION

Chief Mountain Quadrangle, Montana–30 Minute Series
Cordova Quadrangle, Alaska–1:63.360 Series

CONTINENTAL GLACIATION

Jackson Quadrangle, Michigan–15 Minute Series
Palmyra Quadrangle, New York–15 Minute Series
Whitewater Sheet Quadrangle, Wisconsin–15 Minute Series

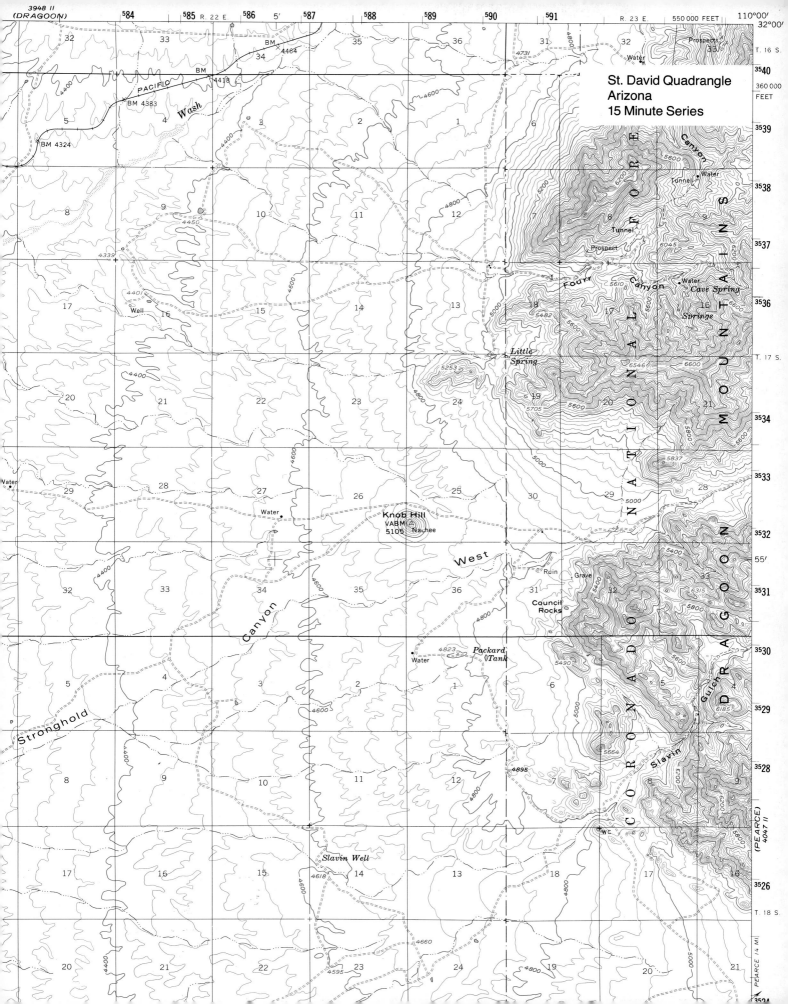

St. David Quadrangle
Arizona
15 Minute Series

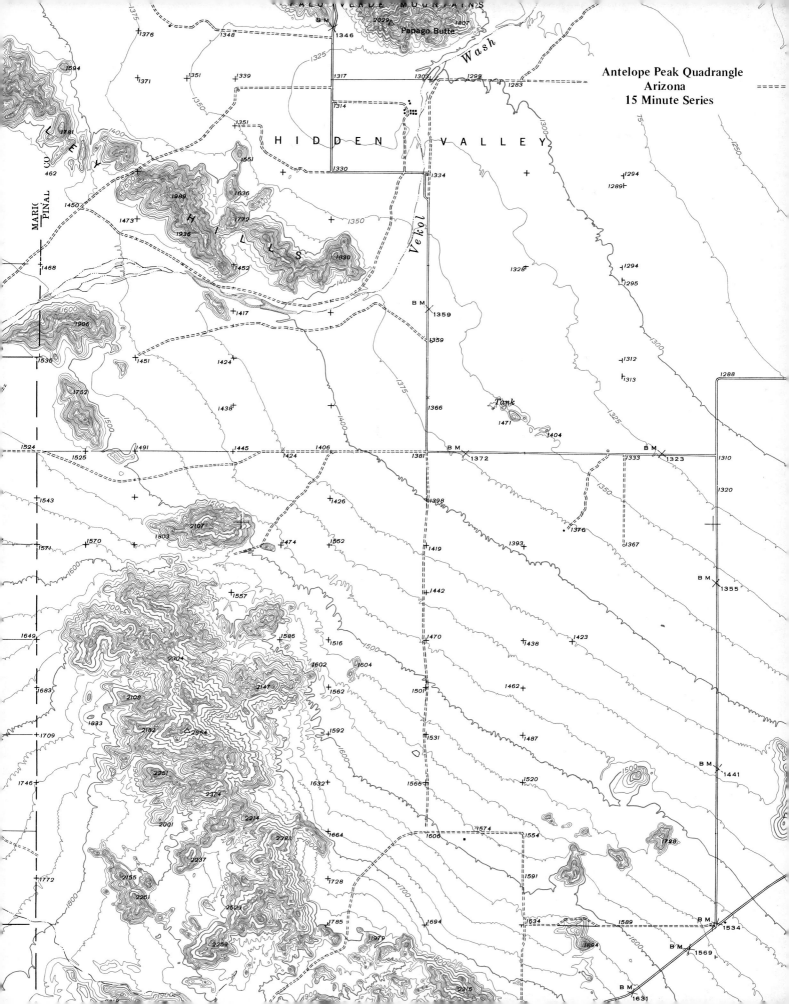

Antelope Peak Quadrangle
Arizona
15 Minute Series

Philipp Quadrangle
Mississippi
15 Minute Series

Ennis Quadrangle
Montana
15 Minute Series

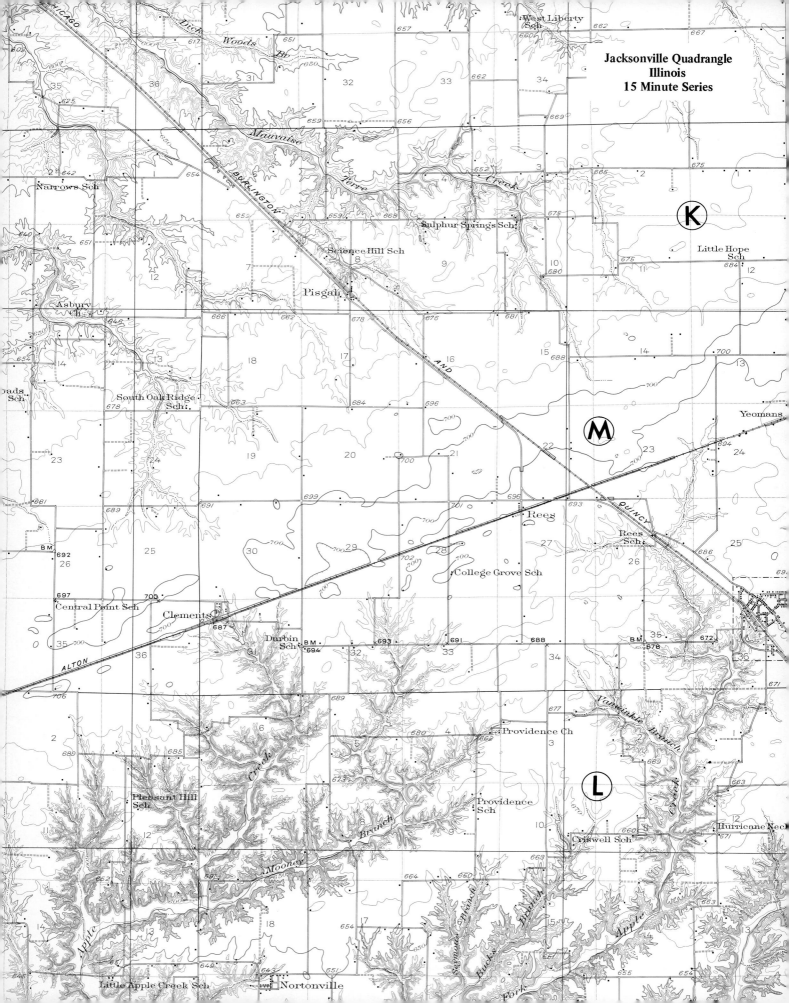

Jacksonville Quadrangle
Illinois
15 Minute Series

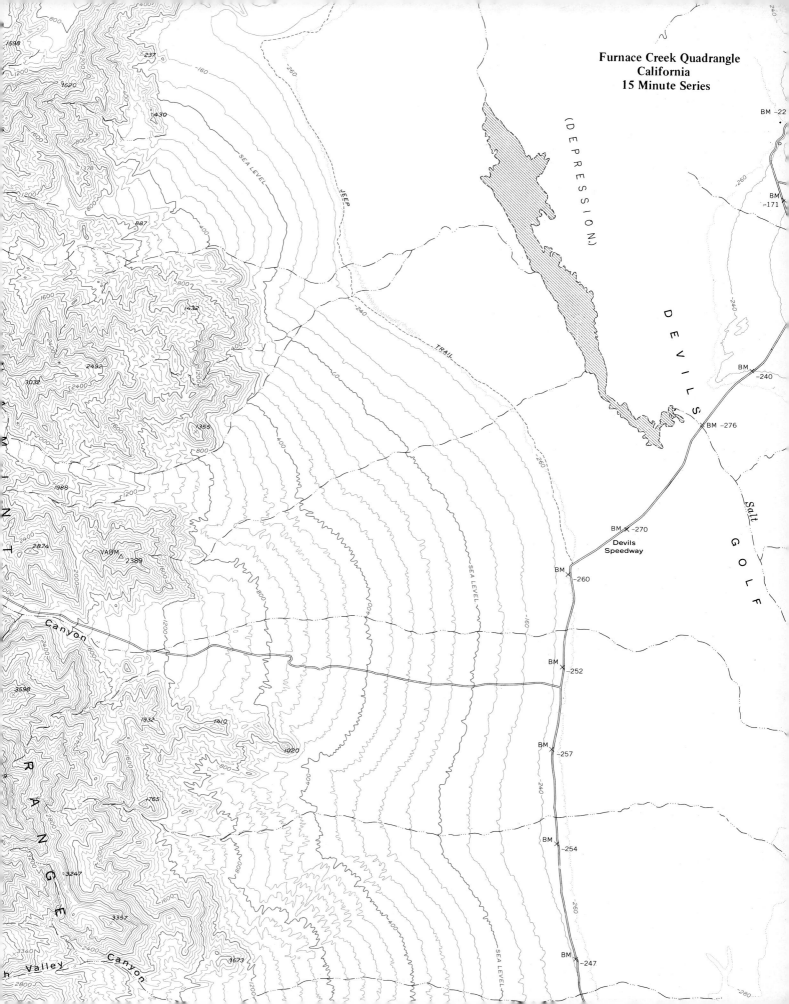

Furnace Creek Quadrangle
California
15 Minute Series

St. Paul Quadrangle
Arkansas
15 Minute Series

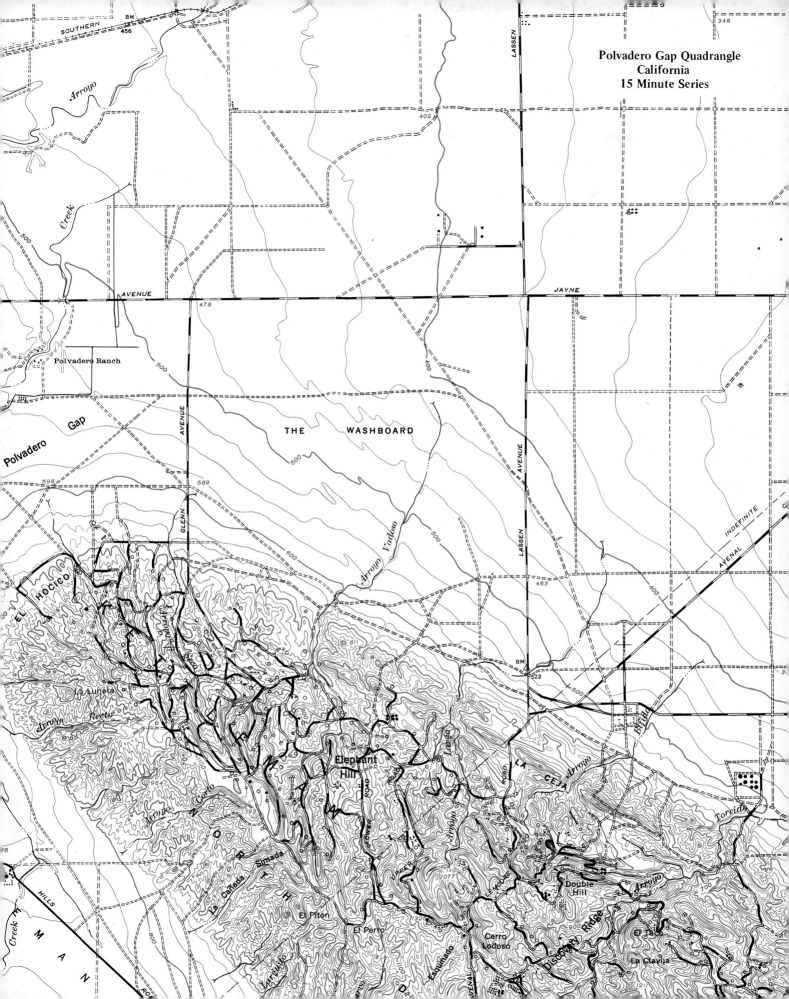

Polvadero Gap Quadrangle
California
15 Minute Series

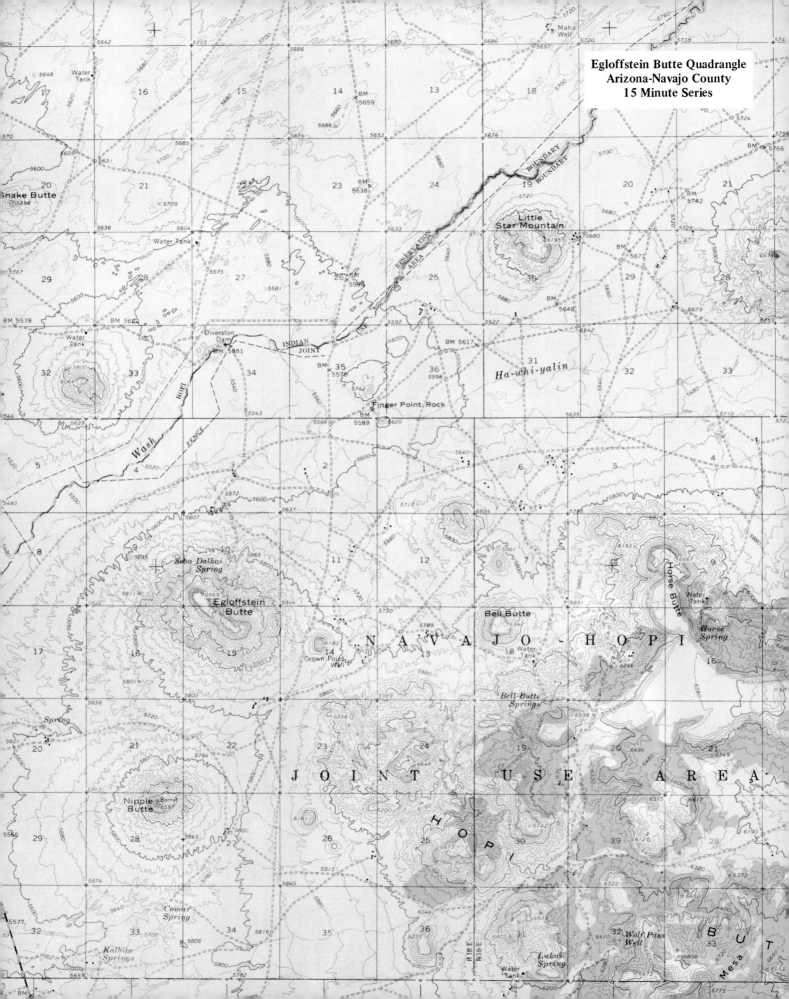

Loveland Quadrangle
Colorado
15 Minute Series

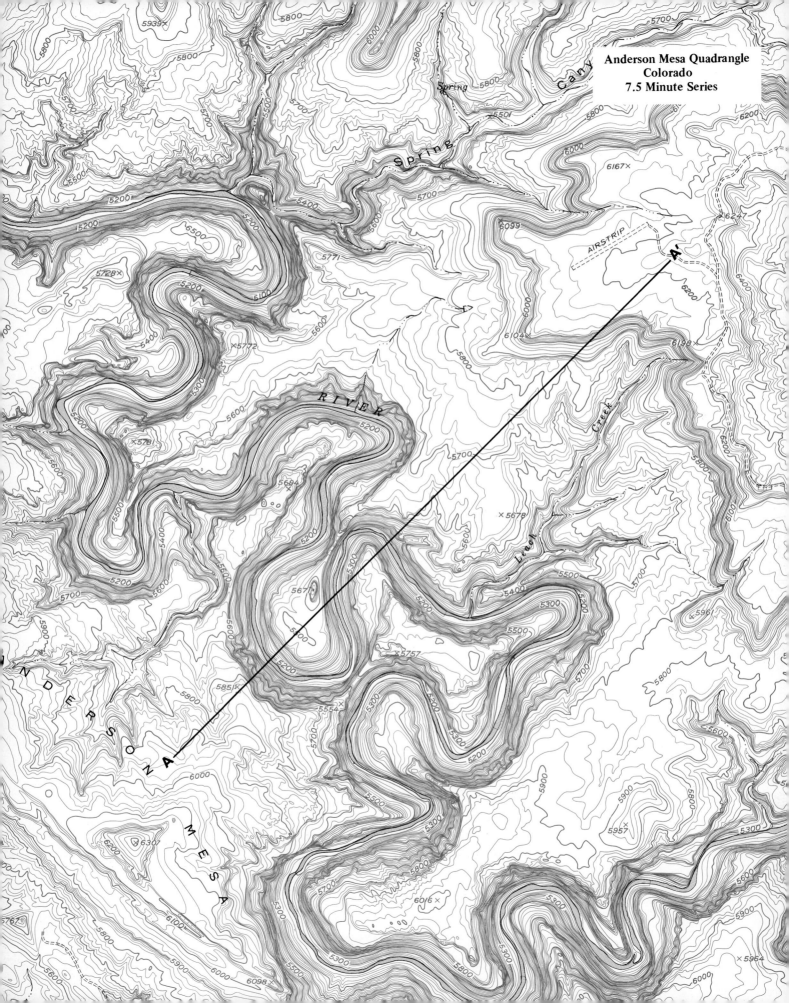

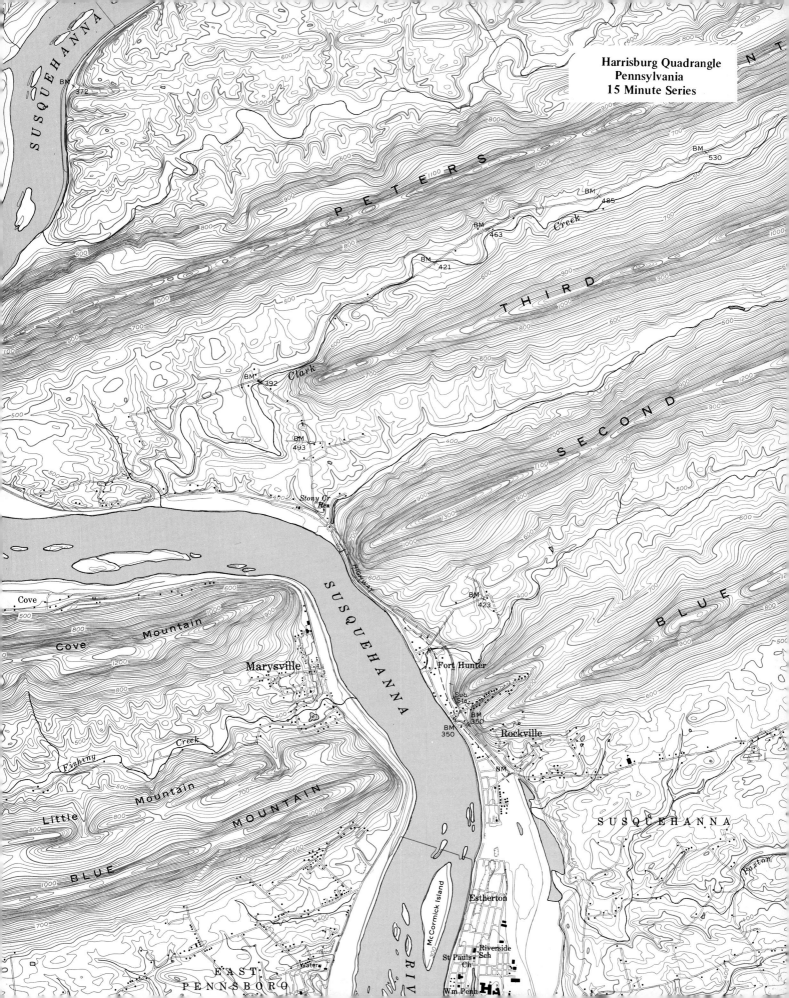

Chief Mountain Quadrangle
Montana
30 Minute Series

Cordova Quadrangle
Alaska
1:63 360 Series   T. 13 S.

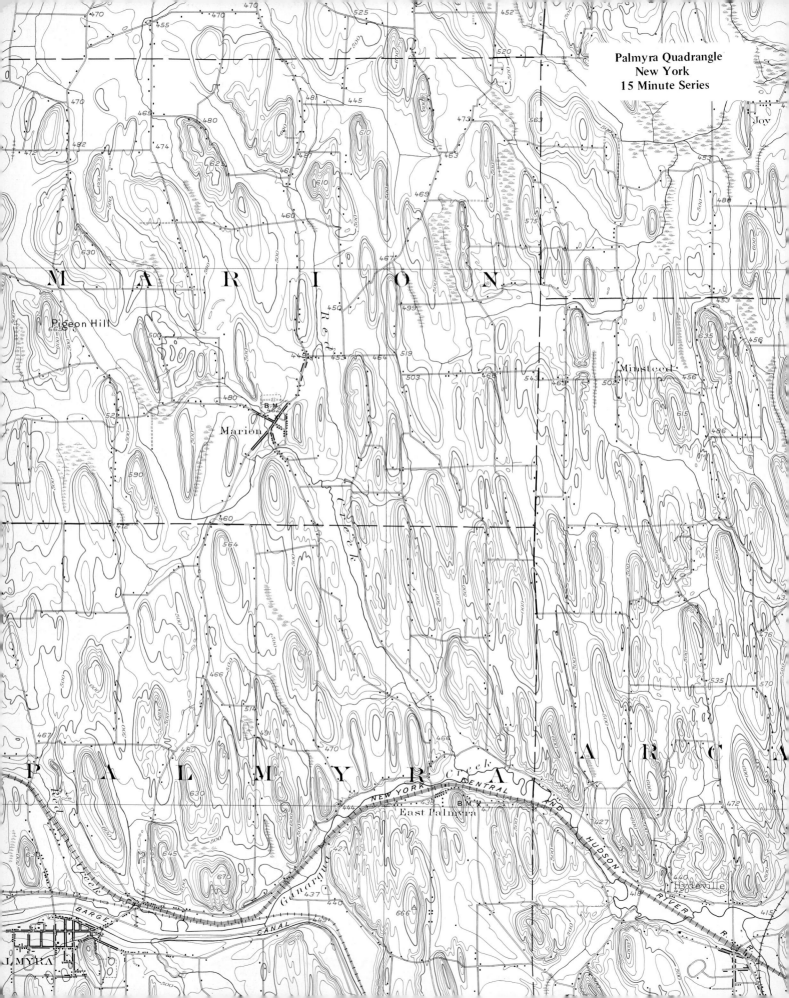

Introduction

A laboratory experience where the sole or main task is memorization, or "cookbook" experiments, or the "plug-and-chug" manipulation of numbers misses what is most essential about the laboratory. Holcomb and Morrison[1] express one purpose of the laboratory very well:

> The essence of learning in science is participation: doing and asking and making errors and learning from them. . . . The lab work is not primarily to train you in certain dexterous actions. It is only to give the genuine feel of that world which we can at best pallidly describe on paper in words or symbols. It is above all to give the sense of what is meant by *abstraction*. . . . we give up completeness by abstraction, but by it we gain knowledge and control.

An introductory laboratory to geology should do two things. The first is to bring firsthand familiarity with minerals and rocks and the ways geologists classify, study, and use them to interpret the earth. It is not genuinely possible to gain a feel for the earth if feldspar, basalt, schist, and breccia are only words or pictures. There must also be that tactile and visual familiarity which comes from handling the specimens, turning them over and over, studying them in ever more detail until the many levels of understanding begin to be revealed. The individual must work with the earth's materials until what is meant by critical observation is understood personally and data collection is revealed as not random or haphazard but as a selective process guided by theories. Study of the earth's materials must continue until the meaning behind classification systems is something not just memorized but understood because personal experience makes the logic of it evident.

The second thing the laboratory should do is give some insight into what is meant by a scientific study and a scientific conclusion. Studies of American students show that their knowledge of science is far below that of students in other industrial nations and has not improved significantly in recent years. Yet we live in a nation and world which is based on and is continuously being altered, for good or ill, by scientific advances. The future of the nation, and the well-being of the enterprise of science, is dependent on a citizenry who understands the process of science well enough to know what is, and what is not, a good scientific decision.

Specifically, you, the student, should come to understand that in science, just feeling, or even "knowing," that something is right is not enough; you must also logically demonstrate the point completely, from facts to conclusions, within a theoretical framework. You must also, with equal facility, be able to demonstrate why another solution is wrong within the theory, and do it in a way that would convince a skeptic. You must come to understand that science is not the random accumulation of knowledge but is the process of using hypotheses to make specific predictions which can be factually and logically demonstrated true, or false.

---

[1] D. F. Holcomb and Philip Morrison, *My Father's Watch: Aspects of the Physical World* (Englewood Cliffs, NJ: Prentice-Hall, 1974), p. 390

The revised sections of this lab manual were written with these goals continuously in mind. We have tried to write exercises that provide many different depths and strategies for exploring earth materials while at the same time being flexible and adaptable. We have included strategies to encourage the student to develop a scientific approach for studying the natural world, and questions and problems that will inculcate a facility for making scientific arguments.

We are aware of how difficult it can be to teach and learn these ideas about science. We are aware of how demanding it can be for the instructor to stay in touch with students who are struggling to learn a new subject and new ways of seeing and thinking. But we believe the earth is an endlessly fascinating laboratory which has much to teach us about the world and ourselves. We hope that the exercises in the manual will make it easier for geologists to use earth materials and earth processes to help students gain powerful insight into the methods, limits, and strengths of science. We also believe that the deeper engagement the students will have with geology through the exercises will more powerfully encourage them to have a lifelong fascination with and appreciation for the earth.

An Instructor's Manual available from the publisher at no charge to laboratory teachers and faculty provides additional strategies and suggestions and, for the first-time instructor, encouragement and hints on how to make a laboratory experience successful for both the instructor and student.